Indigenous Bodies

Indigenous Bodies

Reviewing, Relocating, Reclaiming

Edited by

Jacqueline Fear-Segal

and

Rebecca Tillett

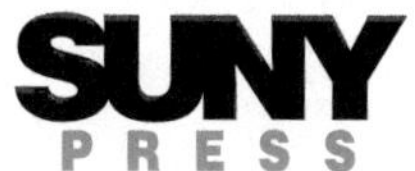

Published by State University of New York Press, Albany

For information, contact State University of New York Press, Albany, NY
www.sunypress.edu

Production by Diane Ganeles
Marketing by Michael Campochiaro

Library of Congress Cataloging-in-Publication Data

Indigenous bodies : reviewing, relocating, reclaiming / edited by
    Jacqueline Fear-Segal and Rebecca Tillett.
        p. cm.
    Includes bibliographical references and index.
    ISBN 978-1-4384-4821-3 (hardcover : alk. paper)
    ISBN 978-1-4384-4820-6 (pbk.: alk. paper)
    1. Indian art—North America.   2. Indian philosophy—North America.   3. Indian
artists—North America.   4. Human body—Symbolic aspects.   5. Human body in art.
6. Human body in literature.   7. Indian literature—North America—History
and criticism.   8. American literature—Indian authors.   I. Fear-Segal, Jacqueline.
II. Tillett, Rebecca.

    E98.A7I48 2013
    704.03'97—dc23   2012045313

10 9 8 7 6 5 4 3 2 1

Contents

Contents

photo gallery follows page 126

Acknowledgments

The editors would like to thank members of the Native Studies Research Network, UK, for their ongoing enthusiasm and intellectual inspiration since the network was founded in 2006. Special thanks go to all committee members past and present: Max Carocci, Graeme Finnie, Mick Gidley, Claudia Haake, Sam Maddra, Jacky Moore, Stephanie Pratt, David Stirrup, Gabriella Treglia, and Robin White. Additional thanks go to our colleagues in the School of American Studies at the University of East Anglia for their untiring encouragement and support.

Chapter 4 is a distilled and abridged version of arguments made by Lynette Russell in her monograph *Roving Mariners: Australian Aboriginal Whalers and Sealers in the Southern Oceans, 1790–1870* (SUNY Press, 2012).

An earlier version of chapter 11 was published in the *Tamkang Review*, "Representing Indigenous Bodies in Epeli Hau'ofa and Syaman Rapongan," 2010, 40.2: 3–19.

Editors' Introduction

Jacqueline Fear-Segal and Rebecca Tillett

When working on the island of New Caledonia, the French missionary and ethnologist, Maurice Leenhardt (1878–1954), in conversation with his most trusted Native informant, Erijisi Boesoou, proclaimed: "In short, what we've brought into your thinking is the notion of *esprit*" [spirit or mind]. To which Boesoou retorted: "Spirit? Bah! We've always known about spirit. What you brought was the body."[1] James Clifford recounts this story in his classic biography of Leenhardt. Outlining how Leenhardt strove to comprehend the different structure of experience that could make such a response possible, Clifford presents Leenhardt's subsequent ethnographic theorizing as a direct, or sometimes indirect, exposition of this rejoinder. He notes how "a dialog of interpretations is portrayed in the anecdote," because it was "an exchange that turns upon Western mind-body dualism and finally unravels it."[2] The story, which has become apocryphal, presents us with a clear reminder of how the concept of "the body" has very specific cultural, historical, and ideological roots. It offers a powerful illustration of how the body's meanings are unstable, open to contest, and can be interpreted differently in contexts where understandings of embodiment are fed from different cultural, or different historical sources. In Paul Rabinow's words, "the intimate linkage between the two key symbolic arenas, 'the body' and 'the person,'" would have to figure prominently on any list of distinctively Western traits.[3]

Acknowledging not only the power of this Western binary but also its dominance and endurance, the chapters in this collection unpack and interrogate this imposed construction, in order to challenge its presuppositions and review, relocate, and reclaim the density and complexity of traditional indigenous beliefs. This book is concerned with the indigenous body as a site of persistent fascination, colonial oppression, and indigenous agency, and the endurance of these legacies within Native communities. At the core of this collection lies a dual commitment to exposing numerous and diverse disempowerments of indigenous peoples, and to recognizing the many varied ways in which these same peoples retained and/or reclaimed agency. Of crucial importance to the contributors to this volume are the ways in which culturally diverse indigenous peoples were—and continue to be—forced to confront and engage with the imposition of the Western mind-body binary as part of a legacy of conquest.

Western concepts of the body have not been static. The earliest split between mind and body derives from the writings of Plato (429–347 BC) and his separation of matter and form. Yet the prevailing modern medical-scientific discourse of the body, as Jonathan Sawday has convincingly argued, has its roots in two distinct but intertwined discursive strands—science and colonization—which simultaneously worked to map the physical body and the larger world, and assert dominion over both. Sawday draws a direct parallel between the dissections and discoveries that were taking place in the famous anatomy theatres opening across Europe—the first in Padua in 1594, followed by Leiden, Bologna and Paris—and European explorations of and encounters with the New World. Eschewing theologically-bound prohibitions on knowledge, he outlines how these scientists accepted no limits on the possibility of gaining understanding. They delved into the inner secrets and recesses of the physical body, which in a previous, church-dominated era, had been considered sacrosanct. "The process," Sawday explains, "was truly colonial, in that it appeared to reproduce the stages of discovery and exploitation which were, at that moment, taking place within the context of the European encounter with the New World."[4] Sawday sees the scientist and the explorer as blood brothers, both intent on knowledge and exploitation, and impatient about notions of mystery and the sacred.

The two interrelated strands of science and colonization came together most fully in the seventeenth century, where the overwhelming impatience to uncover the secrets of the sacred that Sawday describes anticipated René Descartes' profoundly influential scientific philosophy, which separated minds from bodies and atomized each "whole" into its component parts. Emerging as it did alongside the settlement of the "New World" and the development of European empire, Descartes' influence was unprecedented; his infamous cogito of 1637, "I think, therefore I am,"[5] marks the moment when Europe rejected a more holistic worldview and embraced the notion that "the mind . . . is entirely distinct from the body."[6] Such concepts, which were subsequently a central feature of Enlightenment philosophy, had profound implications not only for political thinking about the New World in a burgeoning age of empire, but also its practical application by European colonists; the insistence of the Cartesian scientific "method" that the world—plants, animals, peoples, cultural artefacts—could be broken down, analyzed, and categorized, had an immediate impact in the newly discovered territories.

Descartes' thesis was highly attractive: through correct scientific categorization, "a practical philosophy can be found by which, knowing the power and the effect of fire, water, air, the stars, the heavens and all the other bodies which surround us . . . we might put them . . . to all the uses for which they are appropriate, and thereby make ourselves . . . masters and possessors of nature."[7] Re-inscribing biblical categories that elevated man above the natural world, Descartes' "method" also inscribed new hierarchies between human groups, as subsequent Enlightenment philosophers and scientists detailed the proximity or distance that a variety of human groups had to or from the "natural world." In this context, the greatest impact of Cartesian dualism was in the implementation of a series of ideological hierarchies: minds were elevated (and celebrated) above bodies; human "civilization" and "progress" was ranked by the willingness of social groups to embrace mind-body dualism and by their distance from "nature;" and further physical hierarchies, including those marked by bodily "difference," were justified, facilitated, and imposed, often for economic ends.

The drawing by the Flemish artist Jan van der Straet, depicting Amerigo Vespucci's arrival on the new-found continent (c. 1575), provides a powerful visual allegory of these processes. It was the best-known of many similar images in which the New World was personified as a naked woman wearing only a feather headdress. The focus of much recent scholarly analysis, this portrayal of the contrasting embodiments of Vespucci and "America" carries powerful resonance for this collection.[8]

The image draws clear distinctions between two disparate cultural groups: to the left, Europe, represented by the identifiably patriarchal figure of Vespucci; to the right, America, represented by the figure of a native woman. Yet these distinctions, although initially perceived as cultural, are also political; a recognition of hierarchies of civilization and savagery, and of power and powerlessness. Accordingly, Vespucci stands fully clothed over America's reclining nakedness, surrounded by the symbols of European "civilization": the monetary wealth reflected in his clothing, the scientific and technological instruments that have guided him across the ocean, the military power evident in his ship with its advanced weaponry, and the flag that represents the advanced development of a national identity. By contrast, America's proximity to nature is evident in her lack of fine attire, and in the simplicity of her weaponry; her desire for the progress that Vespucci represents is evident in the delight of her expression as she leans forward to reach for the civilization that he represents. Significantly prefiguring Cartesian dualism, the image behind America of the "cannibal feast" depicts her peoples already actively participating in their own reduction into body parts. Significantly, it is this active indigenous "participation" that continues to be celebrated in Western colonial thinking through the wilful misrepresentation of brutal conquest as mutually beneficial "encounter."

Importantly for this collection, what is striking about this image is the ideological European conflation of indigenous bodies and the lands they inhabit. While it is made evident that America desires European "civilization," America's own representation—her nakedness and her facial expression—suggests a reciprocal European desire for both indigenous bodies and indigenous lands. As Louis Montrose has suggested, this mode of representation acts quite specifically to "sexualiz[e]" the imperial project of "exploration, conquest and settlement."[9] And this sexualization and eroticization of indigenous bodies represents a further development in European ideology that reflects the desires of empire, as complex and nuanced connections between imperial exploration and pornography are made explicit. Matthew H. Edney persuasively argues that clear parallels can be drawn between pornography and imperial cartography "in terms of the objectification of 'other' landscapes and cultures and their subjugation to an empowered imperial vision."[10] Identified variously as "geo-pornography" and "gynocartography,"[11] this recognized patriarchal method of feminizing lands identified as ripe for male conquest also ensured that physical indigenous female bodies could be interpreted and commodified as sites of sexual pleasure and possession; violence and conquest were eroticized and legitimized. From first contact in the Americas, and in Australasia too, colonial articulations of indigeneity were therefore shaped by and contingent on corporeality.

Recent decades have witnessed a wealth of scholarship on the body in a wide range of disciplines, informed by the work of Michel Foucault and his understanding of the intimate relationship between political power and ways in which the body is perceived, managed, and controlled. Two important recent collections have addressed this topic. In 2008, Shino Konishi, Leah Lui-Chivizhe, and Lisa Slater edited a special edition on "Indigenous Bodies," (*Borderlands* e-journal, 7:2), which focused on indigenous Australian issues. A year later, Damian W. Riggs and Barbara Baird edited *The Racial Politics of Bodies, Nations, and Knowledge* (2009), which employs a racial frame to engage specifically with whiteness studies, again within an Australian perspective. *Indigenous Bodies* extends this field in exciting new ways, drawing together the wide range of disciplinary expertise—anthropology, art/ art history, community studies, creative writing, gender studies, health studies, history, literature, political studies, museology, new media technologies, religious studies, visual media—of Native Studies scholars from Australia, Canada, New Zealand, Poland, Taiwan, the United Kingdom, and the United States of America, and emphasizing an international perspective.

The Osage poet and scholar, Carter Revard, has written a foreword to this collection, which stands as both synecdoche and explication of the vast cosmic powers and cultural intricacies that are incorporated and harmonized in indigenous bodies and societies. Revard writes in a wide arc, embracing the stars of the Osage Creation Story, the songs of the storm and of deer mice, and the purposeful activities of tumblebugs rolling cattle dung on the meadows of the Osage Reservation. He relishes and illuminates tiny, obscure, yet profound mysteries that are part of the ebb and flow of life; so his discussion of the humble dung beetle's energetic gathering and shaping of fresh cow manure, is accorded the same eloquence and attention as are the ideas of Ovid and Darwin, deliberately un-doing imposed Western biblical hierarchies and scientific categories in order to emphasize and celebrate an holistic indigenous worldview that not only recognizes but is also organized by symbiotic interspecies relationships. Revard structures his foreword around six of his poems—"Doppelgängers: A Nativity Ode"; "Coyote Tells Why He Sings"; "Æsculapius Unbound" (Ovid and Darwin in Oklahoma); "December Transient"; "Deer Mice Singing Up Parnassus"; "Tumblebuggery"—which he uses to guide the reader toward an appreciation of the profundity and inter-relatedness of all life and, by implication, the trite and damaging reductionism of binary thinking about indigenous bodies.

Revard's foreword immediately establishes the collection's focus on the importance of both indigenous worldviews and indigenous self-representation. It also draws attention to, and refocuses attention on, significant critical developments in the field. Important amongst these is a growing exploration of tribal nationalism and Native sovereignty as scholarly strategies through which to make explicit the otherwise often hazy links between "literature" and "communities," in order to promote and maintain "the intellectual health of Native communities and the quality of critical discourse" within Native Studies.[12] As Jace Weaver, Craig Womack, and Robert Warrior argue in *American Indian Literary Nationalism* (2005), "being a nationalist is a legitimate perspective from which to approach Native American . . . criticism. We believe that such a methodology is not only defensible but that it is also crucial to supporting Native national sovereignty and self-determination, which we see as an important goal of Native American Studies generally."[13]

Indigenous Bodies addresses just such issues. The collection is organized into six parts. Section I addresses visual representations. Both chapters engage with historic representations of the peoples Europeans encountered in their "discovery," explorations, and settlement of new lands. More importantly, they also demonstrate and explore indigenous agency within a colonial relationship that consistently demanded European dominion over indigenous bodies. The widely circulated 1494 illustrated edition of Columbus' letter to

King Ferdinand and Queen Isabella of Spain promised rich lands, gold, spices, and cotton, as well as slaves, and included the first visual images of the New World—the five woodcuts, which supposedly illustrated his voyage. Europe's first sight of the inhabitants of the Americas presented the indigenous population "naked, men and women, as their mothers bore them," signalling a focus on the indigenous body that would have an enduring legacy. Although part of the bounty Columbus offered to Ferdinand and Isabella was indigenous souls, for "turning . . . to our holy faith," more significantly for the purposes of this collection, he also tendered their bodies, to be "slaves, as many as they shall order." From the spoils of his "victory," Columbus believed "all Christendom will have hence refreshment and gain."[14]

Christendom's refreshment and gain demanded America's conquest and domination. In 1992, the Native American artist, Jaune Quick-to-See Smith, chose to mark the Columbian quincentenary by creating *The Red Mean: Self-Portrait,* an art work that tells the story of the damage and destruction wreaked on indigenous bodies and cultures by Europe's "discovery" and invasion of the continent. In Chapter 1, Carolyn Kastner analyzes this work to show how Smith uses her own body to parody Leonardo da Vinci's study of human proportion, *Vitruvian Man,* and, creating a collage from elements of her entwined American and Native American identities, stakes her claim on both American history and Western art history. Kastner continues her examination of the subversive power of Smith's art with *Paper Dolls for the Post-Columbian World with Ensembles Contributed by the United States Government* (1991). This is a single artwork consisting of thirteen paintings that depict narrative moments in the history of the extended contact between the Flathead people and the U.S. government. Three separate paintings of the Plenty Horses family—Barbie, Ken, and their son Bruce—set up a parody of the commercial Barbie Doll, even as it delivers a potent political message. Kastner shows how by creating sets of clothes for their lives away from the reservation, Smith shocks viewers with the enormity of the Plenty Horse family's isolation, loss, and degradation, and the disparity between the two Barbies: Barbie Doll has an infinite wardrobe, while Barbie Plenty Horses has only one dress. The ironic humor of the clothing Smith creates for her paper dolls is essential not only to convey their misfortunes and Smith's political message, but also to the creation of Smith's vibrant artistic narrative, which ensures that the Plenty Horses' story will not be forgotten. Smith insists that she is "never telling stories about 'poor pitiful,'" but "stories about hope with humour," and Kastner illuminates how her art actively reclaims, reconfigures, and reinvigorates stories closed by written history.

When creating her self-portrait, Smith meticulously constructs the complex layers of her multifaceted identity; at all times it is she who fashions and manages how it looks and the messages she wants it to carry. Nineteenth-century indigenous subjects of portraiture, whether paintings or photographs, had no comparable control over the creation of their likenesses. In Chapter 2, Stephanie Pratt presents compelling evidence to show how, nevertheless, some native subjects were able to use their sittings as opportunities to contribute actively to self-expression and self-representation. Analyzing three portraits of distinguished individuals (one of whom is of her great-great-grandmother) Pratt concedes that all these images were made as part of a larger project to provide a record of Native Americans, who were presumed to be in decline and close to extinction. Referencing Mary Louise Pratt's work, Stephanie Pratt goes on to argue that portraits of Native Americans made by painters and photographers from the dominant society can be read as contested sites and should be included as one of the "arts of the contact zone." Stephanie Pratt shows how some objects of material culture, chosen by sitters for inclusion in their portraits, are items of symbolic capital carrying very specific meanings for both them and their communities. By eluding the comprehension of a viewer from outside the culture, such objects are implicitly oppositional; other deliberately selected objects, which are able to be interpreted from within mainstream society, Stephanie Pratt argues, are directly oppositional. While acknowledging the difficulties associated with analysing the significance of each sitter's selected objects, Pratt positions her observations within the growing literature that challenges the colonizer/colonized binary and explores the processes of mediation within cultural products.

The second section engages with the persistent dismemberment and display of indigenous bodies. The frequent and often forcible transformation of indigenous bodies into subjects of entertainment and display was clearly a further result of the colonizer/colonized binary, which fostered a pseudo-scientific fascination with dead as well as living indigenous bodies. In 1882, Phineas Taylor Barnum assembled examples of what he described as "all the uncivilized races in existence," and paraded them round Europe and the United States as a freak show. Barnum's extravaganza provides an illustration of the blurred boundaries between spectacle and science when anthropologists visited his "savages" to take measurements and photographs.[15] Such investigations were widespread in the laboratories and museums of western nations; once deceased, indigenous bodies could be controlled, dismembered, and owned in ways not possible when they were alive. Collected in drawers and boxes and displayed in museum cases and cabinets, indigenous bones and body parts were treated as elements of scientific investigation, not the remains of individuals. They were granted none of the respect normally accorded the dead. For many indigenous peoples, such treatment was not just profane, it was dangerous. As James Riding In (Pawnee) explains, "the acts committed against deceased Indians have had profound, even harmful, effects on the living."[16] The scientific community, triumphant heirs of the Cartesian divide, always viewed the body as the rightful subject of their investigations, and indigenous bodies, deemed to hold the proof of their presumed inferiority, were "docile" and so readily subjected to scrutiny and management, in death as much as in life.[17]

Both chapters in this section on dismemberment and display demonstrate the role played by supposedly neutral scientific investigation in the racialization of indigenous bodies. In Chapter 3, Jacqueline Fear-Segal looks at the complex processes by which supposed racial differences of indigenous peoples were visually constructed and presented for public viewing in the Smithsonian National Museum. Underscoring the public educational function of the museum, she analyzes a single element in the Smithsonian's extensive exhibition—a set of Native American plaster busts that were commissioned by the Smithsonian, using life-masks taken of sixty-four Plains Indians held prisoner in Fort Marion, Florida. Unpacking the power structures that enabled their creation, the motivations for their production, and the manner of the busts' display, she reveals the complex processes by which racial difference (and implied inferiority) of indigenous peoples was categorized, constructed, and exhibited to the public in the nation's capitol. Native American bodies (with special focus on the head) were purposefully included in the Smithsonian ethnology exhibit. The busts' inclusion was crucial, because their presence provided tangible evidence of a racial hierarchy that confirmed the United States' national supremacy. It also legitimated Native American dispossession, explaining and sanitizing the inevitable extinction of America's indigenous peoples.

While Native American extinction was being enthusiastically and scientifically (but erroneously!) advertised and anticipated, in Tasmania it appeared that the recently-arrived settlers had already succeeded in wiping out the local indigenous population. William Lanné, a thirty-two-year-old whaler who died in 1869, was widely believed to have been the "last" Tasmanian Aboriginal man. After his death, his skull was removed and his body was dismembered in the interests of pseudo-science. Some parts were thrown away and, after his entire body had been exhumed, it was ultimately discarded or lost. This macabre episode in colonial history and scientific vandalism has echoed down the ages and continues to haunt the present. In an effort to reverse this process, whereby Lanné the historical figure became Lanné the anthropological specimen, in Chapter 4 Lynette Ryssell reconstructs his life story. She reveals a man who, during sixteen years of whaling, developed competencies and skills that permitted him to transcend the racial categories that might otherwise have restricted him. From the bloodied table of the morgue, William Lanné emerges as someone who was liked and respected with many friends, and who exercised agency over his own labor and daily life. It was only after death, when his body was purloined as a "scientific" specimen to be fragmented, probed, and desecrated, that William Lanné became fully racialized, and co-opted into a divisive racial narrative that his own life had refuted.

Section III explores the exploitation of the indigenous body as a site of sexual pleasure. While there has been much discussion of the male gaze directed at female indigenous bodies, both chapters here engage with the different exploitative piquance associated with the male indigenous body, which is amplified when that body is linked to homosexuality. Drawing upon theoretical ideas that the body is an object and target of power, both chapters discuss how regimes of knowledge produced in colonial situations construct highly specific readings of sexuality.

Drawing on these established discourses of indigenous sexuality, in Chapter 5 Max Carocci traces the ongoing compulsion of the imperial gaze in contemporary gay perceptions and textual representations of indigenous male bodies. Observing the multiplicity of ways in which the male indigenous body has been variously re-imagined, interpreted and articulated in gay cosmopolitan subcultures via texts, pornography, popular music, etc., Carocci argues that the image of the "Indian" emerges as convenient trope that encapsulates many ideas valued by contemporary cosmopolitan gay men. An examination of a long history of perceptions and representations that associate Native Americans with homosexuality and gender deviance, demonstrates an intricate set of relationships between projected and lived realities that identify the male indigenous body variously with sodomy, ambiguity, and feminization. Whilst reflecting on the impact of marginalized groups on the West's own minoritarian constituencies, these intricate relationships invite a reflection on the dialogic relationship between dominant and subaltern positions, in both colonial and postcolonial contexts. Accordingly, Carocci assesses the ways in which American Indians have been represented as emblems of an idyllic and utopian sexual Arcadia, subjects of erotic fantasies, and icons of masculinity, or gender variation. More importantly, Carocci concludes his analysis with a consideration of the ways in which contemporary indigenous gays, lesbians, and transgender people have appropriated/are appropriating and deploying these racist and paternalistic stereotypes to disrupt the ongoing commodification of indigenous bodies in the Western imagination.

In Chapter 6, Murielle Nagy builds on the theoretical structures outlined by Carocci in order to examine the late nineteenth-century work of the Oblate Mission in northwest Canada. Her analysis focuses on one of the missionaries, Émile Petitot (1838–1916), who worked to convert the Inuit, Dene, and Metis peoples, while simultaneously studying the details of their cultures and writing ethnographic texts. Embedded in his ethnographic writing, Nagy reveals a wealth of information about Petitot's own values, emotions, and sexuality, including a relationship with a young, Dene man. Analyzing the many contradictions in Petitot's physical and moral descriptions of indigenous individuals, Nagy uses his life as a means to explore the complexity of the missionary encounter, its problematic relationship to established discourses of indigenous sexuality, and the need for it to be sensitively investigated and evaluated on many levels.

Section IV, on imagination and commodification, addresses the ways in which the fetishization of indigenous bodies as "commodities" throughout the imperial encounter is clearly linked to complex and ongoing colonial discourses. Indigenous bodies are fetishized as savage bodies, enslaved bodies, conquered bodies, sexualized bodies. In the late twentieth and early twenty-first centuries, this fetishization clearly continues in anthropological and legal discourses that aim to identify and evaluate the "authentic" indigenous body; in popular cultural celebrations of indigenous "authenticity;" and in the marketability and profitability of the indigenous body. As both chapters argue, contemporary Native writers are not only engaging with such bodily fetishes, but actively appropriating, challenging, and usurping them.

In a consideration of contemporary Native fiction, in Chapter 7, Joanna Ziarkowska assesses the ways in which writers such as Leslie Marmon Silko, Louise Erdrich, and Sherman Alexie engage with the marketing of indigenous bodies in popular American culture. Identifying the body as a commodity that is susceptible to the laws of demand and supply, Ziarkowska argues that indigenous bodies become desired and desirable "products" because they are culturally "attractive," because they represent colonial fantasies of cultural dominance, and—perhaps most importantly—because they are financially lucrative. However, in order to fulfill such fantasies, indigenous bodies are perceived to require reshaping, enhancing, and improving; in other words, they need to be both "authenticated" and modified by Western culture in order to have "value" (achieve profit) in the cultural exchange. Through an emphasis on physicality and sexuality, and on colonial interpretations of Native "authenticity," indigenous bodies are thus modified and made marketable; while their constituent parts are identified and broken down, enabling a simultaneous appropriation both of the indigenous body and, more problematically, of "indigeneity." Tracing the links between these processes and capitalist gain, Ziarkowska firmly locates such processes within the ongoing legacy of colonial fantasy and its conscious (ab)use to identify the ways in which Native writers are interrogating physical stereotypes in order to demonstrate contemporary Indian realities.

Such contemporary physical realities are also evident in Ewelina Bańka's Chapter 8, which explores the relationships between indigenous bodies and spatial locations in the writing of the Diné (Navajo) poet, Esther G. Belin. Tracing the significance of place and emplacement to Native cultures, Bańka explores the complex interactions between the body and the land, between "home" and identity, between communities and individuals, between families and wider kinship groups, and between maps and mappings of all kinds. Emphasizing the urban Indian experience and the history and legacies of federal urban relocation policies, Belin employs the indigenous body to represent ongoing Native resistance to the colonial fragmentation of Native lands and to situate a reclaiming of both home and indigenous autonomy. Here, the indigenous body graphically represents the impact of ongoing colonial ideology on the urban Indian experience; while the indigenous body itself becomes a site of resistance to racial and sexual discrimination, social marginalization, and cultural estrangement alongside popular appropriation and commodification. Deploying the image of the belly/womb, Belin foregrounds the physical body of the indigenous mother to present a "creation story" and a story of life. This story of "nurture" represents not only Belin's own matrilineal and matrilocal Diné culture, but also acts to affirm the physical survival of indigenous bodies and cultures in the contemporary United States.

Section V explores bodily dis-ease and healing. Regularly viewed as objects of fascination, allure, and desire, indigenous bodies have just as frequently been regarded with disgust and horror, judged to be deviant and requiring control and reform. A drive to reform indigenous bodies (as well as save souls), accompanied almost all government educational policies directed at indigenous peoples across the world. Not just beliefs, but also behavior and outward appearance had to change to match Westerners' notions of the "civilized." Students would be taught to read and write, but they were also required to change their diet, hair styles, and dress. As a result, not all white-educated indigenous people were able to retain a firm sense of identity and pride while undergoing the rigors of a mission or government boarding school education. Many students returned home reluctant or unable to speak their mother tongue, with scant knowledge of their own traditions and culture and a conflicted sense of their identity. And, for some, their education was accompanied by physical and/or sexual abuse.[18] The damage inflicted by these schools is slowly being uncovered. The two chapters in this section analyze the enduring impact of this damage on present-day communities, and examine strategies for healing that reconnect individuals with the beliefs, values, and practices of their traditional cultures.

Suzanne Owen, in Chapter 9, examines how among the Mi'kmaq in eastern Canada, traditional ceremonies, which are accompanied by discomfort and pain, have recently been revived, and other ceremonies involving physical suffering have also been borrowed and adapted from neighboring indigenous peoples. Owens examines the social and psychological significance of these developments. Exploring the interconnections between the experience of physical and psychological pain, she argues that for some Mi'kmaq such ceremonial suffering can provide a means to assuage psychological pain and suffering that is linked to individual and community cultural losses and abuse. In this context, she suggests, physical suffering is regarded as a sacrifice, or gift to "spirit;" when something is asked for—visions, healing—then something must be given in exchange in order to restore the balance. Locating her study within the wider context of the long-term effects of colonialism, Owen uses both interviews and written testimonies to support her argument that the intense physical ordeal of these ceremonies is a way of expressing or expunging suffering that is already present. Prayer with pain in a ceremonial context not only links the individual to community, but can also transform personal suffering into empowerment and healing.

Darrel Manitowabi and Marion Maar, in Chapter 10, extend this exploration of the importance of traditional knowledge and practice to the well-being of indigenous communities. They examine the epidemic rates of type 2 diabetes and its devastating impact on the bodies and lives of indigenous people in Canada, whose over-all health is generally much poorer than that of non-indigenous peoples living with diabetes. In contrast to the physical risk factors linked to this disease in biomedical research, Manitowabi and Maar foreground the indigenous concept of a sense of community well-being (*mnaamodzawin*). They explore how loss of *mnaamodzawin* can be linked to legacies of the colonization process, and is a vital determinant of indigenous health. Analyzing the narratives of eighteen indigenous people living with diabetes on Manitoulin Island, Ontario, their research reveals how cultural, historical, spiritual, and emotional factors impact on both the development and the management of diabetes. This chapter suggests that assessments of the causes and management of diabetes need to embrace historical and political processes *outside* of the body as bearing relational consequences *within* the body, in order to provide alternative ways to improve the understanding and management of type 2 diabetes.

A crucial part of contemporary healing strategies is the relationship between indigenous bodies and the land, which is examined in the final section (VI) of this volume. As the earliest visual representations indicate, in the colonial mind the indigenous body is equated quite simply with the land itself: as Edney comments, indigenous bodies quite literally become "maps" of the newly colonized territories.[19] With a burgeoning European explorative imperial impulse wedded to the newly emerging but increasingly influential Cartesian desire to classify and atomize, the project of mapping "new worlds" became central to the colonial enterprise and, ironically, the indigenous body became central to representations of the "terra nullius." As Jeremy Black argues in his study of *Maps and Politics*, such maps demonstrate "the powerful ability of visual images and messages to represent and advance [political] agendas;"[20] and the long-established coupling of indigenous peoples with the land by the nation-state is evident in a range of academic discourses, including museological representations that display indigenous bodies as "natural history." While such representations foreground the complex contemporary definitions of indigeneity in settler states, and expose the ongoing legal implications of relationships to the land in the context of indigenous sovereignty, they fail to address or recognize powerful and enduring cultural and spiritual relationships to place. More importantly, they fail to recognize the ways in which the long and contested political history of maps and mapping has been reappropriated by contemporary Native groups, to resituate indigenous bodies in their correct relationships with and to the land. The final two chapters in Section VI, Physical Landscapes, address the ways in which two distinct indigenous cultural groups are working to remap their cultural and spiritual connections to the land, both as a process of communal healing and well-being, and as a means by which to promote cultural continuance.

In her analysis of two indigenous Pacific writers, Epeli Hau'ofa and Syaman Rapongan, Hsinya Huang, in Chapter 11, moves beyond established geographic, ecological and archaeological discussions of the South Pacific islands as parables of environmental destruction and cultural collapse, to trace the ways in which indigenous authors celebrate an alternative Pacific where epiphanies, tropes, and regenerations of native attitude are currently taking place. Moving determinedly beyond both recycled nostalgia and misdirected millenarianism, these visions and counter-memories review indigenous performance of bodily techniques and retrieve indigenous body memories. In doing so, they not only reframe macro-geography and micro-politics across the postcolonial trans-Pacific region, but also reclaim indigenous agency. Deploying Hau'ofa's conception of "Our Sea of Islands" (from *We Are the Ocean* (2008)) as a trope, Huang argues that Hau'ofa's work demands a new paradigm for the Pacific to forge a holistic perspective of visionary amplitude and cosmopolitical renewal, through a vast oceanic perception of place, transnational community, islander space, myth, and language. This chapter asserts that Hau'ofa enacts "our sea of islands" as a speech-act of performative and auto-critique, as a way of troping and narrating the nation-leaping expansiveness of the watery Pacific Ocean; and, in so doing, registers counter-memories through an evocative sensibility for "Oceania," with a vision of place and indigenous body in intimate connection.

The collection concludes with an overview of a remarkable digital project to map contemporary physical, cultural, spiritual, and geographic Māori bodies. In Chapter 12, Khyla Russell and Samuel Mann demonstrate the importance and value of a project that makes links between generations, and actively involves younger peoples in the digital remapping of Māori cultural and spiritual space. Conceptualizing a *whānau* (family) or *hapū* (clan) as an epistemological and cultural beginning point, it is made evident that *taha tinana* (a bodily aspect) is part of a wider conceptualization of *takata whenua* (people of the land) and has other aspects which make the whole human. Identifying bodily, psychic, and spiritual connections with *whānau, hapū,* and *Iwi* (tribe/bones), living, yet to be born and dead, Russell and Mann reveal that this is how the Māori connect themselves, through traditional knowledge, to people and landscapes, cultural and geographical. Since land and seascapes are physically connected to through their humanization, the project traces and demonstrates how Māori perceptions of land and seascapes have altered, and have altered them. Creating a profoundly empowering synthesis of old ways of being with those imposed through the processes of colonization, the SimPā Project recalls and reaffirms indigenous storytelling and histories, as part of the bodies of landscapes that the Māori continue to occupy if not own.

The chapters in this collection developed out of presentations and conversations between Native Studies scholars at an international conference, organized by the Native Studies Research Network UK, at the University of East Anglia in 2009: *Indigenous Bodies: Reviewing, Relocating, Reclaiming.* Topics covered in this collection are wide-ranging but not exhaustive: Native art and artists; historical indigenous portraiture; museum display and the dismemberment of indigenous bodies; repatriation; missionary ethnography; gender, sexuality and homosexuality; bodily commodification and mapping; disease and healing; prayer and ceremony; the intimate and profound connections between bodies and lands; and the deployment of new media technologies within traditional cultural practices. Moving scholarly discussion forward in ways that are both innovative and imaginative, the unique international perspective offered here reveals the disturbing ubiquity of the imposition of the mind/body split, and its disastrous implications for indigenous peoples around the world. It also illuminates the creative ways in which indigenous peoples have consistently worked to retain, reclaim, and reinscribe their own cultural values and beliefs.

Notes

1. James Clifford, *Person and Myth: Maurice Leenhardt in the Melanisian* World (Durham NC: Duke University Press, 1982), 172. This story has become integral to scholarly discussions and is referenced in Jonathan Sawday, *The Body Emblazoned: Dissection and the Human Body in Renaissance Culture* (London and New York: Routledge, 1995), 11, and Paul Rabinow, *Essays on the Anthropology of Reason* (Princeton: Princton University Press, 1996), 129.

2. Clifford, *Person and Myth*, 172.

3. Rabinow, *Essays on the Anthropology of Reason*, 129.

4. Sawday, *The Body Emblazoned*, 25.

5. René Descartes, *Discourse on Method and The Meditations* (London: Penguin, 1968), 53.

6. Descartes, *Discourse*, 156.

7. Descartes, *Discourse*, 78.

8. Michel de Certeau, *The Writing of History* (New York: Columbia University Press, 1992), xxv–xxvi; Louis Montrose, "The Work of Gender in the Discourse of Discovery," *Representations*, No. 33, Special Issue: The New World (Winter, 1991), 1–41; Peter Hulme, *Colonial Encounters: Europe and the native Caribbean, 1492–1797* (New York: Methuen) 1986, 1–3.

9. Montrose, "The Work of Gender in the Discourse of Discovery," 2.

10. Matthew H. Edney, "Mapping Empires, Mapping Bodies: Reflections on the Use and Abuse of Cartography." *Treballs de la Societat Catalana de Geografia*, 63, 2007, 91.

11. See Edney, "Mapping Empires," 88.

12. Jace Weaver, Craig S. Womack, and Robert Warrior, *American Indian Literary Nationalism* (Albuquerque: University of New Mexico Press, 2006), xxi.

13. Weaver et al., *American Indian Literary Nationalism*, xx–xxi.

14. Christopher Columbus, *Letter to King Ferdinand of Spain describing the results of the first voyage*, 1493, http://xroads.virginia.edu/~hyper/hns/garden/columbus.html.

15. Roslyn Poignant, *Professional Savages: Captive Lives and Western Spectacle* (New Haven: Yale University Press, 2004).

16. James Riding In, "Repatriation: A Pawnee's Perspective," in Devon Mihesuah, ed., *Repatriation Reader: Who Owns American Indian Remains?* (Lincoln: University of Nebraska Press, 2000), 108.

17. Stephen Jay Gould, *The Mismeasure of Man* (New York: W. W. Norton & Co., 1981).

18. The governments of Australia and Canada (but not yet the United States) have both issued public apologies and the Canadian justice minister described the abuse suffered by First Nations children as "the single most disgraceful, racist and harmful act" in Canadian history. More recently, in the United States, a Roman Catholic religious order in the Northwest, known as the Northwest Jesuits, has agreed to pay $166 million to more than 500 victims of sexual abuse, most of whom are American Indians and Alaska Natives who were abused decades ago at boarding schools and in remote villages. This settlement, with the Oregon Province of the Society of Jesus, is the largest abuse settlement by far from a Catholic religious order, as opposed to a diocese, and it is one of the largest abuse settlements of any kind by the Roman Catholic Church. See William Yardley, "Catholic Order Reaches $166 Million Settlement With Sexual Abuse Victims," *New York Times*, March 25, 2011, http://www.nytimes.com/2011/03/26/us/26jesuits.html.

19. See Edney, "Mapping Empires," 83–104.

20. Jeremy Black, *Maps and Politics* (London: Reaktion Books, 1997), 9.

Of bodies changed to other forms I tell[1]
Tumblebuggery, Creation Stories, and Songs

Carter Revard (Nom-peh-wah-the), Osage

Where We Come From

If you make your bodies of me, you will live to see old age, and live into the Blessed Days. That was what the indigenous beings said to the Osages, when we came down from the stars into this world, and sent our messengers ahead to find out how to live in this new place and time—a Creation Story that used to be told during the traditional Naming Ceremony. The messengers traveled through three valleys—which, the story says, were not valleys—and they met the older beings who had learned to live here. One being they met was very frightening, which caused the Elder Brother, who first met him, to say: *Nom-peh-wah-the!* ("Makes afraid.") But this being, like all those who first met the Osages, was very gracious, and said to them just what they would hear from the others whom they met (Black Bear, Mountain Lion, Golden Eagle, and others): *If you make your bodies of me, you will live to see old age, and live into the Blessed Days.* So the Osages understood that *this* awe-inspiring but gracious being was Thunder, and that the name *Nom-peh-wah-the* could be given someone who belonged to the Thunder clan. (Thus began, moreover, the World Wide Web, which uses the power of Thunder—or, as the Brits would later name it, Electricity.)

In our Creation Story, then, the Osages took the advice of these beings. They *incarnated* their ancient beings and understanding and powers. With songs, and with dances, they *incorporated* the indigenous bodies of these beings into our social structures, shaping our homes and our communities, creating the twenty-four clans of the Osage Nation, over which these great

beings preside as guides, helpers, and name-givers. In an Osage town, the
lodges were arranged in concentric circles, with the lodges of clan members
set in a certain order around the central space of the two great lodges of
War and of Peace, where sanctuary could be had, where food and clothing
could be distributed. When Osages gathered in a "House of Mystery" for
communal actions—a Naming Ceremony for instance—members of each
clan were seated in traditional order in a circle around the central space,
with the Sky half or moiety of the clans on the north side, and with the
Earth and Waters moiety on the south side. The child to be named would
be seated at the East, where the sun rises, and would follow the sun's road
through life.

But the order was deeper yet: in Osage society, marriages had to be
arranged so that one person was from the Sky, the other from the Earth
and Waters moiety, so that every child was born of the marriage of Heaven
and Earth, modeled after the cosmic procreative processes. The assembly of
the Clans—that is, beings who had made their bodies of the great primeval
beings of this world—was therefore an assembly of the beings of Earth, Sky,
and Water; the circular arrangement of ceremonial activities, and of town
plans, was (literally) oriented: social order was therefore meant to reprise the
cosmic order. A child being named was meant to move through life in this
world as the Sun does, rising in mystery and beauty, moving in power and
order, going down in peaceful beauty into the darkness, replaced for a time
by moon and stars, but returning each dawn into this visible world. (One of
our Osage prayer-songs for the morning, which I quote below, says of the
Sun: "He returns, he is coming again into the visible world.")

Osage social order was meant to reflect, or rather to re-create, the cosmic
order, so that everything done was to be as well arranged and sequenced as
the movements of the heavenly and earthly powers and beings. Europeans
may recognize that the Osage world picture of society and government and
natural order much resembles the portrait that used to be presented to us
as the "Elizabethan World Picture," the one whose loss so troubled John
Donne and others. That Elizabethan picture, of course, grew from much
older accounts—those of the Greeks and Romans, those of the Assyrians and
Babylonians and Egyptians before them. Names like Hesiod, Homer, Plato,
Aristotle, Pythagoras, Ptolemy, Plotinus, the Pseudo-Dionysius, and others,
crop up in these accounts.

So, I will discuss some of the indigenous bodies inhabited by our people,
and make a few comparisons—not, I hope, too odious—to the beings and
bodies in the Creation Stories of certain other human societies and nations—
particularly those in Ovid's *Metamorphoses*, and in Milton's *Paradise Lost*,
but also those in the still-being-written *Bible of Charles Darwin*. I will work
outward, both from poems I have written that honor some of the indigenous

beings mentioned in the Osage Ceremonies, and from others who have greeted me in Osage country, that is, the world which is the world of all of us.

And a central focus will be Bodies of Song, because I have come to think no social body is formed without song. I realize all too well that I am not a singer at the drum, and that my poems are not the songs that have made and that keep us Osage, but I will be speaking here as though speech comes directly from song and gestural dance, and I will speak from the belief that without song there are no weddings, by which new beings may come into this old world and sing again. I think even the people whom journalists call "scientists" have begun to perceive that the Darwinian story of natural selection must attend not only to what the eyes see but what the ears hear: not only the peacock's tail but what humans hear as his scream, and the peahen hears as song; and besides, what chicks in the egg, and the nest, hear as mother-tunes, strongheart songs, good advice on choosing life-partners.[2] It seems to me that bird-song at dawn has as much to do with the daily and yearly arrivals and changes of Light as with the arrival of Mating and Nesting Season. I do agree—with the Amen Chorus—that birds begin singing as nesting and mating season arrives, but the question still needs answering: is all song and dance about sex? *Or is it also about harmonizing society (both male/female and individual/group) and expressing joy at the rush of feelings involved?*

I will mention, to support this view, one odd fact: there is a certain moth whose larvae have developed an ability to make their way into the nest of ants who might otherwise devour them, but instead the ants take care of them, feeding and grooming and treating them as royalty—as, indeed, a Queen Ant. And why do they do this? Because the caterpillar or larva sings to them precisely the song that a Queen Ant sings. I suppose only a scientist, or a jazz drummer, would call it a song: the article described it as a series of clicks and twitters, the sort ants make by rubbing or banging parts of their appendages together, something like what we used to do in grade school when I was a proud member of our Rhythm Sticks section, unable to play a harmonica or piano or accordion or anything else. I suppose our drums developed out of such beginnings.

Doppelgängers: A Nativity Ode (if only Columbus had . . .)

It has lately been discovered that, just as the first stanza of this piece narrates, at a certain time of year hellacious gales of wind blow from east to west through certain parts of the Sahara (the "Bodélé Depression"), from which they scoop great quantities of very fine minerals, sweeping them up into dark roiling clouds that are then driven high across the Atlantic, over Brazil and up along the Amazon and its tributaries, where the fine dust eventually settles down into the lush rainforests.[3] In this way, desert and jungle are "Doppelgängers," orchid ("air-plant," epiphyte) an apotheosis of hurricane, nectar an avatar of dust. If we had the ability of angels to see past and present and future simultaneously, we might see jungles that used to cover what is now the Sahara, and perhaps a desert that will cover what are now the rain forests of Brazil; but for now, I have painted only two brothers, African desert and Brazilian rainforest, in present time.

For my poem, I have put that story together with another of an infant's finding his voice, first in weeping and then in laughing, which are also Doppelgängers, and have narrated this in terms of the Osage Creation Story's account of our people's having come down into this world from the stars. So the Infanta Nuova, made of stardust, asleep in a dark house, awakes in pre-dawn darkness and cries, is cleansed, sung to, sings along with the strongheart song, and is fed, then sees through the window the Morning Star and the Dawn, and hears a bird sing, at which (s)he laughs, and sings along with it the new/old song of joy, one of our Osage songs.

In my first year on Earth, my twin sister and I were taken care of for some time by our Ponca aunt Jewell MacDonald in the village at White Eagle, Oklahoma. A lullaby she used to sing us, made by her blind great aunt, is the Strongheart Song she sings in the poem, made to hearten the warriors in despair, driven from their homelands in the Dakotas down to White Eagle in Oklahoma. The old voice is Aunt Jewell's mother, who is waked again at dawn by the child's voice, rises and (like a Ponca Firebird) fixes sun-golden pancakes with honey and fresh butter for breakfast—something gold that sticks to the ribs, a contrafactum to the Frost lyric "Nothing Gold Can Stay." And I have stuffed into the final line words from both Milton's *Lycidas* and the Lord's Prayer.

1.

It's not exactly a Pentecostal wind or
Niña, Pinta, and Santa Maria, it's
just a haboob or maybe simoom, truly
a burning desert blast at this time of the year—
down on the southern Sahara swoops a hellish
roiling hurricane-force wind that scoops
a hundred-mile-long rift-full of dusty crystals up
and up and drives them in dark flashing clouds westward
high and higher and out over the coastline of Africa, the grey
haze now streaming across the Atlantic over Brazil
and on up over the Amazon,
high above lush rain-forests until the fine

dust comes delicately down into an orchid's apotheosis
of hurricane where a hummingbird
glittering sends its long tongue into
deep nectar, avatar
of Sahara sand.

2.

—In this dark house I hear the
shimmering of my Doppelgänger's wings,
but I am crying, the voices say—
some time ago I came down like dust
from the stars into this house where the old voice says
he is crying, give him
some milk, it says,
and the young voice says
I have to *change* him first,
then hands come down and take me up,
remove the swaddling clothes and dip
me in chilly water, wash me clean,
and I am crying and the young voice sings,
I still myself and listen, I hear the words,
"What are you afraid of?" they say,
"No one can go around death."
In this dark house there are no
stars but there is song, the hands
have warmed a bottle, there is milk,
but first I sing along, the young voice stops then
and I sing alone,
"What are we afraid of, no one
can go around death."
My brother hears me and he turns
from the nectar and flies out
into the moonlight, and the stars
are over him. "This child
is singing," the young voice says, and then
the old voice says,
"Give him the bottle, let him sleep."
The milk is sweet and warm. Now
through silent window
the morning star comes nearer,
then fades away, the east turns russet and my brother
the orchard oriole, wearing the soft

colors of early dawn, begins to sing,[4]
so I laugh and sing,
we sing together
without words his song of joy,
"The stars go home and now
the sun appears,"
then the old voice says,
"I guess I better get up
and fix some breakfast now"—
so dawn goes down to day,
its light-gold pancakes lifting off a tray
like little suns, butter and honey spreading,
black coffee's bitter perfume rising while
Grandmother gives us (*yet once more*) our daily lives.

THUNDER: Coyote Tells Why He Sings

We like to think a writer has to find an individual voice. The stress in our academic programs is on *not* sounding like anyone else—as if we were not trying to fool some Queen Ant, or a King of England. I suppose there are classes, film, and drama programs, in which a "new Shakespeare" could learn and be taught something about writing plays for a group of actors, and I suppose that, in our time, those classes must all try to make sure we could recognize any play he wrote as by this "Shakespeare." Speaking for myself as one small writer, thinking about the fragile greatness of Imperial America, I have tried to recall how I found a voice to speak in this great wilderness. Where I found it was in Oklahoma, Land of the Red People, and it was a certain indigenous being, called by the Aztecs Coyotl, who gave this voice, so on public occasions it seems good to read out the words in which he first showed me how the sounds of the world turn into music:

Coyote Tells Why He Sings
There was a little rill of water, near the den,
That showed a trickle, all the dry summer
When I was born. One night in late August, it rained—
The Thunder waked us. Drops came crashing down
In dust, on stiff blackjack leaves, on lichened rocks,
And the rain came in a pelting rush down over the hill,
Wind blew wet into our cave as I heard the sounds
Of leaf-drip, rustling of soggy branches in gusts of wind.

And then the rill's tune changed—I heard a rock drop
That set new ripples gurgling, in a lower key.
Where the new ripples were, I drank, next morning,
Fresh muddy water that set my teeth on edge.
I thought how delicate that rock's poise was and how
The storm made music, when it changed my world.

When I was writing the poem, thirteen lines came along very easily and quickly. I realized that it would need just one more line to make a sonnet, but had to wait several hours before that line dawned on me, on Indian Time. It was only by waiting that I found what the poem was about, that is, what the story within the poem was telling me: *the storm made music, when it changed my world*. Thunder created song.

Meeting Some Other Indigenous Beings

We move now into three other offshoots of creation stories, one told by Ovid, another by Charles Darwin, and a third by whoever wrote the Book of Acts in the Christian New Testament—stories that in my poem *Æsculapius Unbound (Ovid and Darwin in Oklahoma)* take part in and metamorphose the poem's central story, which first grew out of the Oklahoma Dust Bowl and tells of the killing of two egg-stealing snakes by a cowboy who had married an Osage woman. In the poem, the boy who watched the snake-killing later heard, one summer twilight, the evening song of an orchard oriole. This story couple-twines with one told by Ovid in his *Metamorphoses*, Book Fifteen—one of the last tales in that long work, after he has told and re-told how the world was created, and endlessly changes and interacts.

The story Ovid tells is how Apollo's son Æsculapius, god of healing,[5] changed himself into the great serpent that twines about his staff, and in serpent form was brought by the Romans from the ancient Greek shrine in Epidaurus to Rome to be installed, as the God of Healing, in a new shrine on an island in the Tiber, where he healed the Romans of a terrible plague. The second main Creation Story within my poem was composed by Charles Darwin and is re-told in many encyclopedias under the entry for *Reptiles*. This story says birds and snakes evolved as branches of *Reptilia*, and that mammals bloomed on another branch of that tree: so, in Darwin's myth, we humans (including the Oklahoma cowboy who shot the egg-stealing snakes of the poem, and the boy who witnessed that and heard an oriole sing) also hang, not far away from snakes and birds, on this tree of life. In the Darwinian epic, our kinship with birds and snakes appears in the form of the three smallest bones in the human body—those of our inner ear, which allow birds and humans the miracle of hearing, but in snakes, which have no sense of hearing, are the three bones that allow them to unhinge their jaws and swallow things bigger around than they are. And the third story (which comes in at the very end of my poem when the boy is trying to re-create the song of the orchard oriole by whistling it) is told in "Acts of the Apostles" 2:1–6.

Aesculapius Unbound
(Ovid and Darwin in Oklahoma)
By beautiful design, a snake's jawbones
unhinge, so it can swallow
things bigger around
than it is.
I wondered, when
the old man shot a blacksnake in
the hen-house, then held it
up by the tail,
just how in hell those great big lumps
along its six-foot length, slow-twisting up and
down as it hung, had ever been
choked down.
Later, I heard that snakes
are deaf, those three hinged bones
had not yet turned into the malleus, incus
and stapes of my middle ear,
they have no tympanum
and no cochlea,
no auditory nerve, their brain only
processes earth's vibrations, but not thunder's,
so snakes don't sing, although perhaps
they dance when mating—
only their cousins, the small birds,
sing the light's changes,
as Melampous knew.
So when I heard,
as twilight grew, the orchard
oriole sing its heart out there in the elm's
Edenic shadows, something unhinged
and let the music in,
but if they hold me up and listen,

it may by then be part of me—
be how I live and breathe,
or will at least be how
I try to whistle, when
the spirit moves.

December Transients

To human ears, a great many birds don't sing, and we have other verbs to refer to the noises they make. I was impressed lately with an accurate description that somebody offered in a poem, of the noises made by migrating geese as they fly overhead: this poet spoke of the sound as like that made by pulling rusty nails out of wood. In the poem you are about to read, which tells of hearing geese on a December evening in St. Louis, I describe the sound as "creaking honks." No doubt other geese hear it as more like the clarion sound of Birgit Nilsson in, say, *The Flying Dutchman*. But the indigenous bodies within my poem are those of migrant geese passing over St. Louis, and of a father out for an evening walk with his son and the son's dog Blanca, an almost all-black mixture of Border Collie and Mutt, rescued from a pound. Tweeting away within this story, however, is one that I boldfaced in with two lines pinched from Shelley's "To a Skylark"—the story of a bird that does sing, according to human ears, very beautifully indeed. And once again, the Creation Story told by Charles Darwin re-shapes the St. Louis narrative at the poem's center.

December Transients
First snow four inches deep, a half
moon white gold as twilight darkens and Blanca—
off-leash, racing in circles, panting with joy—hurls herself
down and rolls, and snuffs, and wriggles into new
half-fluff, half-sleet gleaming beneath
dark trunks and leafless branches in
the little park, leaps up and grins
trotting back to Lawrence, who cocks his head
and looks up saying, "Geese!"
as the creaking honks come faintly down
from darkness, and overhead appear, spot-lit
in the golden lightning of the sunken sun
and soaring moon, a ragged V of geese out of darkness
floating, shifting and milling overhead,

uncertain whether to pause, and then
they reform as three move to the lead, the lines
tighten and they head southwestward into
the afterglow and are gone.
Who ever could
have thought, watching the small
dinosaurs shiver while the first
brilliant snowflakes came down into
their world so newly cold, that a small
black Border Collie would take such joy
in running through that snow
with featherless biped friends, while dinosaur cousins
with feathers, calling to the moon and sun, would soar
above them brilliant as snowflakes, on their way
into that world of living waters toward which,
this night, they fly away.

One reason I have boldfaced the lines taken from Shelley, **in the golden lightning/of the sunken sun**, is that whenever I have showed a printout of this poem to friends, the first thing they have said is how beautiful those lines are. At that point I felt a bit like the Keepers of the British Museum: not willing to give up the Elgin Marbles, but compelled to acknowledge that they are "borrowed." So I have put a bold face on my theft from Shelley.

Deer Mice Singing Up Parnassus

On earth, at every hour of day and night, something is singing to celebrate the marvel of mundane revolution. Turn now to one small nocturnal singer, the white-footed mouse or deer-mouse. I won't get into the fuss over whether these are one species, or how many subspecies and variants there may be: if you Google up "Deer Mice" you can read for yourself, and you can also see photos and hear recordings of them singing—both the California and the Eastern Deer Mouse, or White-Footed Mouse. For now, I will say only that years ago when I was teaching an Honors section of freshman English, an essay I asked the students to read with me was one by the marvelous "nature-writer" Sally Carrighar, in which she described hearing what seemed a bird trilling, then saw it was a deer mouse. My friend Bill Winchester tells me that when deer mice came into his house from the tallgrass prairie of Oklahoma, he live-trapped and released them in a nearby hedgerow, but they waltzed back in, singing an epithalamium.

As for other stories within this story about singing deer mice, the first thing to say is that the word for *Muse* in Latin is *Musa*, in Greek *Mousa*, in English *Muse*; and the English word *Mouse* is linked to all these by an *O*—which I assume gives a singing deer mouse poetic license to party on Parnassus and drink from springs where the Muses dance. The gallant cavalier mouse Reepicheep (from the *Narnia* stories), and that immortal trio from *Twelfth Night* Sir Toby Belch, Sir Andrew Aguecheek, and Feste, grilling Malvolio about Pythagorean metempsychoses—all can join the deer mice in a catch, a coranto, a galliard, jig, or sink-a-pace, singing to moon and stars, Diana and Venus. And still another story, lightly revised, fits into this festival: Blake's "Sunflower," seeking that sweet golden clime where the Traveler's journey is done, is a new song-home for deer mice, who came before tourists with FOX2P genes arrived.[6] And yet more stories dance in, of John Muir and his unsuccessful fight to preserve a pristine valley called *Hetch Hetchy* in the Sierra Nevada: the dam was built despite Muir's efforts, and now the people of San Francisco drink vintage water, piped down to them from that reservoir. I doubt not that St. Francis, for whom their city is named, delights in deer-mice.

But of course there are two other stories within this poem: the first is the Osage Creation Story of our coming down from the stars, and the second is one told by John Milton in his Latin epigram written for the great Neapolitan singer Leonora Baroni, whom he heard in Rome in 1637 or 1638.[7] Milton's point is that when Leonora sings, what he hears is above the simply angelic music that we may hear, the music that is mediated to us by our Guardian Angel or *Genius*—the intellectual soul we are born with that guides us and in whose "voice" we can "hear" the musical harmony of the universe. Leonora's voice, Milton says, is that of *Psyche*, the divine Intelligence of the Third Heaven; and when she sings, that divinity "glides through her throat" and through our ears into our minds/souls, letting us participate in the higher harmony of the universe, which Music can make audible to us in such a performance. It is a higher version of that deepest harmony in the old Ptolemaic "Music of the Spheres," which comes from the angels who cause each Sphere to rotate, as each of these angels utters its single note and all the notes join in a "diapason" or heavenly harmonic chord.

These Neo-Platonic notions are relevant once we ask the simple question: where does that little deer-mouse's song come from? He was singing when our Osage people first came into his world, and he goes on singing, just as our people do, to keep on living in this world. Plotinus, Ficino, Milton: exalted company, of course, for a Country Mouse, but I have added as bodyguards for him a sabre-tooth tiger from the La Brea tar pits, and a salmon from the Pacific Ocean swimming back up into the Sierras, and a rainbow trout rising after Mayflies—fit company for Il Piccolo Pavarotti as he sings his way up Parnassus to dance with the Muses.

Deer Mice Singing up Parnassus
(for Bill and Lois Winchester)

In this "new" world they sing

      as we come down from the stars,

  like Milton's Leonora singing

   (aut Deus, aut vacui certe mens tertia cœli),

    they climb up the stems

   of sunflowers still not weary

       of time, and they trill,

     perching and swinging,

   in meadow and glade, as if

      a rainbow

    trout might rise

      to Mayflies from their

    music, as if John Muir and

    Hetch Hetchy

      might come back

   alive and listening,

      anadromous as salmon or sabretooth

  tigers, up time itself into the glistening

     moonlit sonatas of

      Sierra song.

Tumblebuggery

Let me move on now to a poem written back in the days of Bush Two, when we were being told by Paul Krugman and others that there was an economic disaster ahead, but when Wall Street and the City of London and other wiseacres were still in denial. It struck me how very like global bankers were to the tumblebugs I watched, in our bluestem meadow during the Dust Bowl. The meadow was prime grazing land, where cattle of course deposited a great deal of "manure." Back then, we went barefoot from April through September, and our meadow was prime roaming space: the hay in June was up to our knees at least, and once feet were toughened we ran through it at will, raking bare feet on prickly vines of flowers and weeds and watching for the occasional copperhead or rattlesnake coiled and waiting. Meadowlarks, bobwhites, scissortails, goldfinches and dickcissels, kingbirds, barnswallows and purple martins, flew and sang and called around us. Butterflies, bumblebees, passed from flower to flower, pausing to sip and pollinate. Now and then a skunk had to be politely passed, and our house and barn cats, out on their own expeditions, might be murdering some of the meadow's birds and rabbits.

At dawn, and in the twilight, we might have to go up to the far end of the meadow and bring in our Jersey cow to be milked, though usually she would come when called. Sometimes, we'd kneel and watch for a fascinated ten minutes the dark shiny tumblebug beetles—digging out their perfectly round holes, scurrying back to a pile of fresh cow manure and gathering some into a neat globe, turning and using the hind feet to roll the globe along to the hole and pushing it down, going in after it and staying for a time, then coming out and going back for another globe.

So, many years later, somehow the word *Lucre* brought those tumblebugs back to mind. I began remembering them, even as I was reading a passage in *Piers Plowman* about the Seven Deadly Sins, and in particular the Sin of Covetousness. And when I opened the *New York Times* at the breakfast table, and read another column by Paul Krugman, and thought some more about the bankers, and the unbelievable greed and display of wealth taking place around the world even while the wealthy were crushing the poor in the "undeveloped countries," from whom much of their wealth was taken, it struck me that these bankers and money-gatherers were behaving remarkably like tumblebugs, though hardly so harmless, and not always as useful.

So I began writing about tumblebugs in Oklahoma: and I had been thinking also about what *Money* comes down to, which is *Credit*, and of course the root of *Credit* is "belief." I was thinking that when you trace it through, all money is a gigantic Ponzi scheme, in which you have to believe in bankers, and bankers have to take your money and "make use" of it, which means they have to *believe* in the people to whom they lend it out "at interest"—and *Interest* means *That Which Is In Between*. "Interest" is a legal euphemism that replaced the morally disapproving term *Usury* in the sixteenth century, hotting up capitalism. I had not so long ago re-read Herman Melville's *Confidence Man*, and realized it was very much about these matters, and that he wrote it when the American banking and currency systems were being developed.[8] And I had long been delighted and amazed by Alexander Pope's *Third Moral Essay* on the use of riches, especially the couplet in which he describes one result of changing from gold coins to paper money: "Blest paper credit, last and best supply,/That lends corruption lighter wings to fly." Pope lived, of course, before electronic transfers of "money" were possible. And naturally I had long since learned the etymologies of *Money* and *Mint*, which go back to the Roman temple of the goddess *Moneta*, "she who warns," in which temple I think the Romans had their coining machinery and operations.

But as I wrote about the Oklahoma tumblebugs, I got curious as to what a more Darwinian account of them might tell me, so I looked them up in the *Encyclopaedia Britannica*. The Brits don't really know how to speak American, so there was no entry for *Tumblebug*. Eventually I found "dung-beetle." And then to my surprise I learned that Oklahoma tumblebugs are

very closely related to Egyptian scarab beetles, which did with camel-dung in
the desert what tumblebugs did with cow-manure in the Oklahoma dustbowl.
Whole new vistas opened from that: think of all the marble "scarabs" that link
Egypt to Greece and Rome, think of the Egyptians placing a scarab beetle on
the breast of a deceased Pharaoh as symbol of immortality.[9]

Then it occurred to me that on the back of the American dollar bill
there is a *pyramid*, with a shining eye at its apex. This led to my looking at
the Great Seal that is also on the back of a dollar bill, and its history—which
taught me that the motto on that Seal was taken from the *Aeneid*, showing
that even in 1782, before our American Constitution was being framed to
proclaim this a new Republic, the motto on our Great Seal was taken from the
great poem of the Roman Empire. *Wikipedia* says the Great Seal was created
in 1782 by commission from the Continental Congress, explaining the motto
thus: "The Eye over it [the pyramid] and the motto *Annuit Coeptis* allude to
the many signal interpositions of providence in favor of the American cause.
Annuit Coeptis is translated as *"He (God) has favored our undertakings."*
Then *Wikipedia* adds: *"Annuit Coeptis* and . . . *Novus Ordo Seclorum,"* can
both be traced to lines by the Roman poet Virgil. *Annuit Coeptis* comes from
the *Aeneid*, book IX, line 625, which reads, *"Iuppiter omnipotens, audacibus
adnue coeptis."* It is a prayer by Ascanius, the son of Aeneas: *"Jupiter Almighty,
favor [my] daring undertakings."*

I began to see into American history a little better, because I was
trying to look closely at Oklahoma tumblebugs. And when I showed a draft
of the poem to my friend and colleague Carl Conrad, who taught Classics for
many years at Washington University, Carl reminded me of something that I
afterwards put into a footnote to the poem: Aeschylus, in a choral ode from
Agamemnon, refers to Ares, God of War, as "money-changer of corpses." YES!
I thought, money, global banking, tumblebuggery, they do fit together—and
the God of War watches over them all.

As you will see, there are a number of stories in this poem. It was hard
to find a way to end a piece where so many stories were jostling each other,
so I decided to make the ending more fantastic and funny. Whether it works
well enough, perhaps another year of re-thinking and revision may help me
learn. Meantime, I hope it makes a certain amount of sense as it now stands.

Tumblebuggery: O Dea Moneta, Credit Our Belief
(Oklahoma and Wall Street, 2008)

In-CRED-ible, what those brilliant
tumblebugs would do, out in our deep
bluestem meadow there on the Osage
Reservation, where placid cows in August daily and nightly

deposited their sloppy green
hendiadys, treasure and feast. Who knew,
when we would step with care,
barefoot as we were, around
those piles of green cowshit as we walked, on our way
to bring the cows back home and milk them,
how those dark-suited Scarab Beetles
assumed the role of global bankers, working
to "speed the process of converting
manure to substances usable
by other organisms"?[10] As barefoot kids, of course,
in Dust Bowl Oklahoma, we knew bankers
not at all, nor had we yet looked closely at
the back side of George Washington (on such green
money as poor kids had, a dollar bill),
where it looked back at us, that shining eye
of all-foreseeing Providence
inside the Great Seal of the United States, atop
its Pyramid, encircled by that pious, that august,
that Empire-founding invocation borrowed from
Ascanius, son of Æneas,
Annuit Coeptis Novus Ordo Seclorum.
And even when we kids knelt and watched a shiny
tumblebug rolling, rolling, turning upside
down and pushing with her hind feet onward
the lucreous globe of shit she'd made,
and we saw her pause,
and scurry round to where
she'd dug the hole just big enough, measuring with
antennæ twiddling,
then shove and shove the ball down and in,
and go in after—I think we didn't know
that she'd be laying eggs just where
each hatchling's mouth would find a feast,
before she came back out, retraced her trail and started
to shape another global deal almost
worthy of Ares, money-changer of dead bodies,
or Melville's *Confidence Man.*[11]
We laughed at her solemn scurrying, all
klutzy but precise—although the more we looked,
the more we saw how purposeful she was.[12] Yet even if
our one-room school had owned

an old encyclopedia, we'd surely not have
looked under "dung beetle"
to see what we were seeing: that wasn't
what we called them, let alone
"Scarab Beetles"—so none of us, scrabbling there
in our August Dust Bowl, would have looked through a
Britannica, or an Insect Bible, back into
that old Egyptian Empire where embalmers placed
on Amenhotep's breast a Sacred Scarab, the
Tumblebug of Immortality that signified
the Sun rolling the dark ball of Earth and with
"the 30 segments of its six legs [standing for]
the 30 days of each month"—[13]
but even if we had, I think some smart twice-born kid
(Dionysian, Christian, or In-di-genous?)
might just have said: no wonder Pharaoh gave
Columbus so little credit that he left
the Pyramids behind and shook the dust
of Egypt from his feet and emigrated from
that ancient Dust Bowl to our green
and Promissory Land of Con-fidential
and Pyramidal Credit,
America.

Notes

1. Ovid, *Metamorphoses*, trans. A. D. Melville (Oxford: Oxford World Classics, 1987) p. 1, I. 1.

2. See, for instance, Tim Birkhead, *The Wisdom of Birds, an Illustrated History of Ornithology* (London: Bloomsbury, 2008)—especially Chapters 5, "Light and the Breeding Cycle"; 7, "Choristers of the Groves, Birdsong"; and "A Delicate Balance, Sex." It is a perfectly marvelous book.

3. For scientific accounts of this, see *Deflation in the dustiest place on Earth: The Bodélé Depression, Chad*, in *Geomorphology*, vol. 105, Issues 1–2, 1 April 2009, 50–58; and *Proceedings of the National Academy of Sciences of the United States of America*, December 8, 2009, vol. 106, no. 49, 20564–20571.

4. One year orchard orioles nested in the elms beside our home, and I learned to whistle their challenge-notes and the long cascading series of mellifluous notes of their song. Alexander F. Skutch (*Orioles, Blackbirds, and their Kin*, University of Arizona Press, 1996), studied them in their winter migration homes in Central America and says the orchard orioles were "most songful of all the birds I have heard. . . . At dawn, young and old sang together in a many-voiced chorus of whistled notes delightful to hear" (190).

5. It will be recalled that Socrates, after drinking the hemlock that would cure him of living, requested that Crito sacrifice a cock to Æsculapius. Socratic irony, indeed.

6. See *New York Times*, 29 May 2009, A5: human "language gene" put into mice deepens their baby-cries.

7. You can read that epigram, and a fine English translation of it by Lawrence Revard, in the new Blackwell's edition by Stella Revard of *John Milton, The Complete Shorter Poems* (p. 199).

8. As for Melville's *Confidence Man*, gene studies identify his father as Moby Dick, his son as a Robber Baron, and most Congressmen and Presidents as descendants.

9. For a relevant discussion, see Ingrid D. Rowland's "The Charms of Ancient Egypt," in *The New York Review of Books* LVI, No. 11 (July 2–15, 2009) 17–8, reviewing *Les Portes du Ciel: Visions du monde dans l'Egypte ancienne* (a catalog of an exhibition at the Louvre), and *L'Egypte ancienne entre mémoire et sciences*, by Jan Assmann (Paris: Hazan/Musée du Louvre Editions).

10. *Encyclopædia Britannica: Micropædia III* (Chicago, 1983), *s.v. dung beetle*, p. 709.

11. Aeschylus, in a choral ode from *Agamemnon*, refers to Ares, God of War, as "money-changer of corpses": *ho chrysamoibos d' Ares somaton*. Some people think the Trojan War was a global deal.

12. See John Perkins, *Confessions of an Economic Hit Man* (Penguin, 2006), *passim*. He claims to have been trained as undercover operative to ruin the economies of countries in the Middle East and South America on behalf of U.S. and international corporations collaborating with U.S. intelligence and governmental managers for this purpose, and tells in considerable detail how this was planned and carried out in order to gain control, and profit from, wrecking the economies of countries while pretending to help them. His book should be read in the context of what is happening right now in Peru, where U.S. and other foreign oil interests are dictating the destruction of Peruvian forests, rivers, and the indigenous peoples who live in and by those woods and waters.

13. *Encyclopædia Britannica, Micropædia III*, p. 709.

I

VISUAL REPRESENTATIONS

Jaune Quick-to-See Smith

Indigenous Bodies, Indigenous Stories in a Post-Columbian World

C**AROLYN** K**ASTNER**

My work comes from a visceral place—deep, deep—as though my roots extend beyond the soles of my feet into sacred soils. Can I take these feelings and attach them to the passerby? To my dying breath, and my last tube of burnt sienna, I will try.

—Jaune Quick-to-See Smith[1]

You don't have anything / if you don't have the stories. Their evil is mighty / but it can't stand up to our stories. So they try to destroy the stories / let the stories be confused or forgotten.

—Leslie Marmon Silko[2]

American artist Jaune Quick-to-See Smith is a storyteller, but she doesn't fit into tidy categories of art or life. She invents her artistic strategies to suit her immediate goal, which is always to attract and inform an audience. Her visual vocabulary and materials are as diverse and expressive as her interests. Smith is a bricoleur, who has self-consciously shaped her identity and her art from disparate cultures and ideologies. In 1992, she directed her artwork to an American audience familiar with Columbian celebrations of discovery and conquest. As an educator and political activist, her intention was to compli-cate that celebration by introducing a counter-narrative that personalized the devastation of contact and the methods by which the rich indigenous cultures of the American continent were impoverished by colonial conquest, domination, and assimilation. Smith generated a new narrative in several works of art in two series entitled *I See Red* and *Quincentenary Non-Celebration*. In addition, during that same year, Smith curated an exhibition of thirty-four Native artists entitled *The Submuloc Show/Columbus Wohs: A Visual Commentary on the Columbus Quincentennial from the Perspective of America's First People*.[3] The artist re-imagined the long-standing Columbian celebration by visualizing indigenous bodies that survived to tell their own stories.

In *The Red Mean: Self-Portrait* (Figure 1, photo gallery) from the 1992 series *Quincentenary Non-Celebration*, the artist stakes her personal claim on American history and Western art history. The specific story of Jaune Quick-to-See Smith's identity is literally inscribed on her body outlined in black, an indexical mark traced by her husband as she lay on a support of newspaper. Entitled *The Red Mean*, it resonates with the classical ideal of the golden mean, even as "red" asserts the artist's cultural identity in opposition to the European lineage of art traced to ancient Greece. Smith grounds her art and identity in the cultural and temporal present by drawing on pages of the *Char-Koosta News*, the official paper of the Flathead Reservation of the confederated Salish and Kootenai tribes, where she was born in 1940.[4] The artist makes news, as her body confronts the viewer on the ninety by sixty-inch canvas. She positions herself at the center of the front page just below the masthead, as she reaches her arm across the page entitled "Local News." Continuing to build her narrative, she collaged the declaration "made in the USA" across the chest. Below that she marked the body with her stenciled enrollment number, 7137. The number identifies her as a member of a federally recognized Indian Tribe.[5] The collaged and stenciled text are signifiers most often constructed as opposites—American or Indian. Here, Smith is explicit. She is as American as she is Indian. The viewer is on notice that nothing is as simple as it seems in Smith's artwork—she is telling a new story.

Jaune Quick-to-See Smith's enrollment number places her precisely in a matrix of United States government regulations and records, a colonial system set in place during the nineteenth century that continues to define who will be recognized as an Indian by the federal government.[6] Further, it places the artwork within the discourse of contemporary Native American art, where artists often refer to their indigenous blood quantum or enrollment numbers as a visual strategy to complicate perceptions of sovereignty.[7] The complexity of Smith's narrative fragments and signifiers, fabricated from found materials, also relates the work to the practice of American pop art. Smith is explicit on this point. "I appropriate pop art because it is symbolic of American mainstream culture. This gives me a common language in order to communicate with the viewer."[8] Smith's unequivocal tactics identify her simultaneously within American art and Native American art.

Just as significant as the declaration of tribal affiliation and appropriation of American pop art is Smith's strategy to situate herself in Western art history. She challenges Leonardo da Vinci's Vitruvian man from 1490 by encircling the outline of her outstretched body. The contour of Smith's body breaks beyond the delimiting circle that seeks to contain her form, as she dares to transform and exceed Leonardo's man, while breaking free from the weight of patriarchy and Western art history. The red circle takes on layers of meaning, as a medicine wheel, target, or the crosshairs of a gun site, aimed at the figure's abdomen. In art, as in her life, Jaune Quick-to-See Smith challenges every limit by building from all of the experiences that create her exceptional identity as an American artist. Her style and medium are firmly planted in Western European traditions of painting and drawing. Even so, she signifies difference by crossing boundaries to form a unique visual vocabulary weighted with her cultural history. This artwork demonstrates Smith's belief that: "There is a particular richness to speaking two languages and finding a vision common to both."[9] Her artistic and political power is achieved in her compositional balance between the discourse of modernity and indigenous wisdom. It is a dynamic equilibrium that she has been recalibrating in each new series, since she began graduate school in 1978.[10]

Jaune Quick-to-See Smith is a member of a generation of Native American writers and artists, born in the first half of the twentieth century, who preserve traditional stories even as they articulate the discourse of modernity. With concentrated discipline they have mapped unique forms created to bridge opposing worldviews. For example, N. Scott Momaday's *House Made of Dawn*, published in 1969, is a modernist novel, with serious responsibilities in the tradition of Native storytelling.[11] The first sentence is a single word. "Dypaloh." A marker of difference, the traditional invocation of Jemez storytellers locates the narrative within a cultural tradition unknown to the majority of his readers, and co-locates the novel between mythic and linear time. Similarly, Leslie Marmon Silko's *Ceremony*, published in 1977, is structured simultaneously as a novel and as an indigenous ceremony. She shapes a mixed-blood protagonist whose hybrid cure is predicated on his ability to construct a coherent identity from cultural fragments. Any reader may participate in the novel's cure, appropriate to modern alienation. These writers transcend cultural traditions to express their personal experience.

In their anthology *Reinventing the Enemy's Language*, Joy Harjo and her collaborators named this process by which contemporary Indian writers continue to serve their communities as storytellers.[12] It is by using this concept of "reinventing the enemy's language," that the accomplishments of these modernists become legible. This generation of writers embraced the challenge of expressing communal values and life outside of linear time by translating these ideals into English in the structure of the modern novel. Similarly, indigenous artists have invented new and imaginative forms to expand the practice of American art. The artwork of Jaune Quick-to-See Smith, Fritz Scholder (1937–2005), T. C. Cannon (1946–1978), Kay WalkingStick (1935–), and Helen Hardin (1943–1984), to name only a few, expresses the power of this position. Each artist has imagined a new visual language by seizing and deploying specific signifiers to create new meaning in multiple layers.

Jaune Quick-to-See Smith's visual narrative, *Paper Dolls for a Post-Columbian World with Ensembles Contributed by the U.S. Government* (Figure 2, photo gallery), is a particularly powerful example of "speaking two languages and finding a vision common to both."[13] Created in 1992, it is set against the quincentenary. It is a single work of art shaped by thirteen watercolor drawings.[14]

The artist drew the outline of each figure and several changes of clothing on eleven by seventeen-inch sheets of paper and then painted the details in a watercolor wash. Painted with the immediacy of political poster art, Smith devised this multi-part work to be easily portable. Unframed, the paintings can be easily stacked and mailed, which facilitated its wide travel to oppose the quincentennial narrative of Columbus between 1992 and 1994. Assembled and installed in a gallery, *Paper Dolls for a Post-Colombian World* is a significant work of art that occupies thirty-four hundred square inches with a storyline that builds from image to image. Smith created each painting as a narrative moment in the history of the extended contact between the Flathead people of the Bitterroot Valley of Montana and the forces of the United States government and the Catholic Church.

The title, *Paper Dolls for a Post-Colombian World*, immediately identifies the political and cultural space of Smith's story, calling attention to the particular status of Native peoples living within the United State, five hundred years after first contact. Like her enrollment number in *The Red Mean*, the phrase "post-Columbian" foregrounds the colonial relationship of power endured by the indigenous peoples of North America. Contrary to the very notion of "postcolonial," Smith identifies the continuing colonial position of the indigenous peoples living under federal government regulation. The title signifies the unrelenting colonial structure that persists to classify the legal status of Native peoples by blood quantum in the post-Columbian world.[15] This artwork adds Smith's voice to the thousands of indigenous people, who challenge the legal and economic structure of colonial rule. Her semantic play on the philosophical marker "postcolonial," which is frequently used by scholars to critically analyze life after colonial rule, brings attention to her acute awareness of contemporary discursive theory. More importantly, the artist focuses the viewer's consciousness on the political position circumscribed for indigenous peoples in the United States and ruptures the discourse of "postcolonial." Smith's entire career as an artist, curator, lecturer, and political activist, has been directed toward disrupting the colonial discourse, in a search for an exchange of ideas in a shared language that expresses the multiple cultures living in the post-Columbian world of the United States.

Paper Dolls for a Post-Columbian World is the particular story of the Plenty Horses family and their dislocation under colonial power. Like her self-portrait, Smith personalizes the story of colonial domination. However, while *The Red Mean* is a significant statement of difference and survival, *Paper Dolls for a Post-Columbian World* is an extended and more elaborate narrative mapping the specific process of cultural transformation for a Flathead family. Smith's story is pointedly articulated from the subject position of the Plenty Horses family, complicating the euphemistic label "assimilation" posited by colonial supremacy. This shifts the narrative for the viewer and expands the possibilities for multiple narratives. By fashioning various "ensembles contributed by the U.S. government," the artist invites each viewer to create new meanings from the layers of representation, metaphor and text. The cumulative exchange between artist and viewer creates an intimacy that invokes the experience shared between traditional storytellers and their listeners, who hear an enlightening narrative told from memory by an individual. Jaune Quick-to-See Smith reinvigorates the storytelling tradition of her father, Arthur Albert Smith, in the realm of contemporary art. Like her father, the artist tells a story of Salish-speaking people that will not be easily confused or forgotten. Smith has fond memories of her father.[16] She remembers him as a man raised in the old ways by his grandmothers and aunt, and as a wonderful storyteller, who spoke Salish.[17] However, because she was forbidden to speak her father's language as she was growing up, the artist preserves a traditional storyteller's point of view in the enemy's language.[18] From her personal position, Smith builds a story of post-Columbian experience.

Her narrative begins with paintings of Barbie and Ken Plenty Horses and their son Bruce, as a knowing critique of the dolls designed by Ruth Handler in 1959. The artist appropriated familiar American cultural symbols, the Barbie doll and her longstanding boyfriend Ken, to tell the unfamiliar history of the Flathead people of Montana. As Smith parodies such well-known figures, she reaches a broad audience with humor. The doll, known simply as "Barbie," is located in a matrix of twentieth-century American consumer culture, which she enjoys with Ken.[19] Jaune Quick-to-See Smith was not the first to parody these dolls, however, she contributes to the discourse by reinventing the enemy's language. She speaks fluently in the visual language of Western art history, in the dialect of explicit reservation humor. Suspended between these two traditions, Smith's Native humor binds and transforms these disparate cultures to deliver her subversive political message. In the tradition of earlier pop artists, Smith alludes to popular culture specifically to charge her figures with a critique of the American consumer culture that has supplanted Native traditions.[20]

The familiarity and humor generated by the pop culture references belie the serious intent of Smith's narrative. Through the personal lens of one family, Smith tells the story of Chief Charlo and the Salish-speaking people, who were forced from their ancestral home in the Bitterroot Valley in 1891.[21] The artist imagines Barbie Plenty Horses and her family as trickster figures, those familiar and beloved characters of North American Indian stories who entertain and teach. Irony is an important part of trickster action. The trickster delivers a lesson by shifting viewpoints and ideologies, as opposites are bound together. Smith's intention is radical in action and disruptive to the dominant social and cultural values.[22] With this work of art, Smith opens a dialogue closed by written history. She affirms her position: "I am always turning myths upside down and backwards."[23] "I defend, affirm, hold up, the other realities the true ones, the human ones."[24]

Theorist Gerald Vizenor interprets this kind of humor as a sign of survivance, the expression of tragic wisdom that endures as a source of trickster humor.[25] Jaune Quick-to-See Smith survived, to tell her family's story of tragic wisdom with humor. She declares: "Reservation humor derives out of hardship, and survival. Reservation jokes are often bleak and self-deprecating, which is true of any community of color. We laugh at our own shortcomings."[26] Smith provokes humor in a tragic story by transforming pop culture icons into the oppositional indigenous protagonists of her narrative: Barbie, Ken and Bruce Plenty Horses. She performs a trickster shift, as she layers humor and popular culture with the history of the forced assimilation of the Flathead people to create a new meaning for her audience.[27]

The dislocation of the Plenty Horses family from the land is immediately apparent in the context of Smith's larger body of work. The absence of landscape in *Paper Dolls for a Post-Columbian World* visually segregates it from the majority of Smith's artwork. Before Smith received her graduate degree in 1980, she was already exhibiting and selling her inhabited landscapes in New York City. Her *Wallowa Waterhole* series of 1978 (Figure 3, photo gallery) is an example. Named for the verdant valley in Oregon, where Nez Perce Chief Joseph and his people lived and bred their appaloosa horses, each lyrical pastel forms a memory map. Coded with color and line to signify waterholes and fields of wildflowers, the landscape is marked by bird tracks, and dotted lines denoting the trails of humans and horses. These paintings recall the pre-reservation idyllic life of the Nez Perce and share a graphic language with ledger drawings of the nineteenth century. They represent the landscape of Native imagination, an endless expanse of land traversed at will by animals and people.

Ten years after her romantic evocation of the Wallowa waterhole, Jaune Quick-to-See Smith's artwork no longer referenced the romance of a lost paradise, even though she continued to deploy images of animals, plants, and humans across her abstract canvases.[28] As her artistic practice matured, her landscapes most often addressed the conflict of contemporary political issues, but she continued to lead with aesthetics. As always, her intention was to project a subtlety that attracted viewers with color, composition, or the grace of a line. Smith describes her strategy: "My work has layered meanings, so you can see many different levels within one single work. I like to bring the viewer in with a seductive texture, a beautiful drawing, then let them have one of the messages."[29] It is only as viewers approach the art that they realize Smith's landscapes present a history lesson or an ecological message, embedded within the dense layers of paint or held in place by a tracery of lines. It is all part of her plan, to speak in the enemy's language of aesthetics to deliver her message.

In contrast to her landscape style of edge-to-edge painting, Jaune Quick-to-See Smith isolates each member of the Plenty Horses family in *Paper Dolls for a Post-Columbian World*. Visually and metaphorically, she ruptures their communal way of life. Each member of the Plenty Horses family stands alone in an unpainted indeterminate space on his or her separate scrap of earth.[30] Individually embodied, framed discretely, they stand severed from each other, the earth, and their nurturing community. Before she even begins to dress these paper dolls in their new clothes, Jaune Quick-to-See Smith has already signified the greatest loss experienced by Native peoples in forced assimilation—the loss of homeland, the source of ritual, and community.

Smith's handwritten English text is another signifier of separation and distance. The narrative of each panel is in the explicit language of the enemy and leaves no doubt about the artist's message. The surname "Plenty Horses" is a traditional name that denotes wealth and status among the Salish people.[31] As an English translation from Salish, it announces the assimilation process has already begun.[32] Though Smith never represents the United States government with a figure, the ferocious English text on each panel articulates the language and position of the government. These small details of land and language may be overlooked at first glance, but they are the pivotal signifiers of Smith's shifting, oppositional, and indigenous viewpoint.

Jaune Quick-to-See Smith extends the narrative of the Plenty Horses family by creating ensembles that tell a more legible and accessible part of her story. In opposition to the endless wardrobe of the Barbie doll, the artist creates only one dress for Barbie Plenty Horses. This is in sharp contrast to the Barbie doll, whose extravagant wardrobe implies her leisure status. Barbie's identity as an incessant shopper for clothes to be worn in upper-class pursuits such as riding, swimming, and dancing at balls, parallels the incessant demand by the Barbie doll's owners for these new outfits. The comparison to Barbie Plenty Horses couldn't be more severe, as Jaune Quick-to-See Smith implies the restricted existence of a woman who needs only one outfit—a maid's uniform—"For cleaning houses of white people / After good education at Jesuit school."[33]

Comic action can't happen in isolation, it must be inclusive.[34] We smile at Smith's trickster shift because we share knowledge of the Barbie doll's popularity and the attributes that distance her from indigenous culture. This dissonant pairing of the two "Barbies" relies on the surprise connection.[35] Gerald Vizenor describes this kind of indigenous humor as the defining characteristic of contemporary Native literature, the element that provides a work with a sense of cultural truth.[36] Smith's focus on the clash of cultures shocks viewers with the disparity between the two Barbies. Confronted with the colliding worlds, viewers are disoriented. Drawn closer, they are primed for a sustained moment of contemplation and Smith's reinterpretation of the historical text of assimilation. Engaged by the humor, the trickster lesson is revealed.

Reifying the displacement of the viewer's expectations, Smith repeats the gesture of a limited wardrobe by creating just one ensemble for Bruce Plenty Horses, who shares the boarding school experience with his mother. She makes it obvious that Bruce's identity has been so contaminated that a simple change of clothes cannot express it. To articulate his position, the artist creates a discrete second Bruce to visualize the changes wrought by his boarding school experience.[37] Standing in his new haircut and clothes, his demeanor is eternally sad and, even more dramatically, his skin color is changed. Bruce's narrative is described in the enemy's language of English text that menaces the child from every direction. "After being taken from Parents because gov't wants Indians to assimilate. After education child will grow up and get a good job as day laborer. Haircut and white people's clothes, manners and religion will make a Happy and Healthy child. Priests will whip if child speaks Salish."

Smith's meticulous transformation in the appearance of Bruce Plenty Horses is not an act of poetic license; it is based in her knowledge of the countless before and after photographs that were taken at boarding schools from the East Coast to California. Used by officials as evidence of their success in transforming the students at these schools, the pictures also document the student experience of fear and degradation. John Choate created the most famous photographs of Native students at the Carlisle school in Pennsylvania. The most famous pair of these before/after photographs pictures Tom Torlino (Navajo) before (1882) and after (1884) forced assimilation with significant changes in his skin color, haircut, and clothing.[38] He was a student at the Carlisle school in Pennsylvania, far from his family and home in the American Southwest.[39] Devastatingly, the damage is as palpable in those painful photographs as it is in Smith's rendering of Bruce.

Unlike Barbie and Bruce, Ken Plenty Horses has many ensembles that mark his changing role in his family and community. Smith creates multiple ensembles for Ken to explicitly trace the dislocation of the Plenty Horses fami-ly.[40] He leaves his home still dressed with pride in a capote. Smith's beautifully rendered image reminds viewers of the loving community expressed in Ken's clothing, made for him by a sister, his mother, or his wife. The wrap, made to keep the horseman warm as he hunted to feed his family, is now the dress of exile, "For traveling on Forced Removal After Garfield Treaty 1891."[41]

Like his namesake, the Ken doll, Ken Plenty Horses has no responsibili-ties. Imprisoned and forced onto reservations, the role and status of Native men were pointedly and particularly diminished.[42] The life they had known ended abruptly and for countless numbers, so did their self-respect and pride. As Jaune Quick-to-See Smith astutely suggests in Ken's clothing, it was an incremental process of stripping away their dignity in a series of demean-ing exchanges that diminished their role and status. Life-affirming customs were disrupted as the cultural patrimony of ritual and ceremony was severed from traditional communities. The artist visually isolates the headdress from the warrior and declares its monetary value in the hands of a collector, as a semantic shift that articulates a corresponding shift in power. Headdresses, like moccasins, shields, and clothing, were specifically made for warriors. Each item communicated a particular narrative of bravery, generosity, or respon-sibility. Collected and worn by non-Natives, cultural patrimony signifies the status of the collector, as the stories of its origin are confused and forgotten. As portraits of non-Natives dressed in hybrid costumes suggest, they gained status from trading and collecting among the Indians.[43] The same was not true for the Native peoples, whose patrimony was stolen, confiscated by govern-ment officials, or sold during periods of great economic and cultural duress.[44]

Smith signifies the abrupt and humiliating end to Ken Plenty Horses' position in his family and community with the specific wardrobe she creates for him. His clothes trace the decline of a warrior and hunter removed from his homeland, forced to live under United States government occupation on a reservation. Smith expresses Ken's ruin and separation from his former life in his "special outfit / For Trading Land with the U.S. government / For whiskey with Gunpowder in it." Dressed in the "Suit For Receiving U.S. gov't rations," the heroic horseman and hunter wears commercial clothing appropriate to the degrading act of being dispatched periodically to receive government hand-outs to feed his family. This is a familiar story among the tribes that were forced onto reservations. Across North America, diverse cul-tures came to share a common history and a common enemy.

Jaune Quick-to-See Smith articulates that common story as she shares her personal narrative.[45] Belgian Jesuit priests founded Saint Ignatius Mission on the Flathead Reservation, where Smith was born. Jesuit Father le de Ville's white face identifies him as the only non-Native among the paper dolls. The artist names and dresses him as an exotic figure compared to the Plenty Horses family. She visualizes the Jesuit priest as an accomplice to the United States government, noted in the title as the contributor of the ensembles for the Plenty Horses family. As teachers and missionaries, the Jesuits enforced English language and Christian religion, which opposed the traditional culture and belief systems of their parishioners. Father le de Ville represents the face of the institutional forces that severed ties among family members and ultimately desecrated their traditional community.[46]

The final degradation for this family and thousands of others is signified in the outfits imagined by Smith capable of fitting anyone. The seriousness of "matching smallpox suits" and the hospital gown that "fits all members of an Indian family" distends the comfort zone of humor and the meaning of "ensembles contributed by the U.S. government." The specter of the faceless figures covered with spots mercilessly confronts the viewer. The final chapter in Jaune Quick-to-See Smith's story of visual delight and wit is delivered in this deadly serious history lesson.

The artist's narrative might have easily fallen into a bleak story of nostalgia or melancholy. "But I am never telling stories about 'poor pitiful,' I am telling stories about hope with humor," declares Smith.[47] By telling her story of tragic wisdom through the appropriated identity of the Barbie doll, the artist generates a healing humor as viewers imagine a new idea about the human experience of the Plenty Horses family. Her visual strategy invokes the tradition of the solace that heard stories impart to listeners.[48] Though her artwork speaks in the language of the enemy, her viewers participate in the experience of a trickster lesson that entertains as it teaches. As Gerald Vizenor reminds us: ". . . [T]he cruelties of civilization had dispossessed the tribes of their land, but not their stories. The shadows of their words, the hermeneutics of survivance, are forever true on the land stolen from the tribes."[49] Jaune Quick-to-See Smith transforms American art, popular culture, and history, as she creates indigenous subjects generated by narratives, expressed as survivors who live to challenge written history.[50]

Jaune Quick-to-See Smith's work, "in response to the United States Government's Quincentennial Jubilee or the 500-year anniversary of the landing of Christopher Columbus in the Caribbean," is a living trace of her cultural survival, despite assimilation.[51] The artwork and exhibition created by Smith during 1992 is a record of her triumph as an artist and curator, who expands the boundaries of American art. More importantly, each time a viewer pauses to make meaning from her art, Smith's story is remembered. The very life of her artwork assures that Jaune Quick-to-See Smith's story will not be confused or forgotten as it continues to reinvigorate Salish storytelling traditions, in the language of American contemporary art.

Notes

1. Lucy Lippard, "Jaune Quick-to-See Smith's Public Art: Generosity with an Edge," in *Subversions/Affirmations: Jaune Quick-to-See Smith*, Jersey City Museum exhibition catalogue (Jersey City, NJ: 1996), 80.

2. Leslie Marmon Silko, *Ceremony* (New York: Viking Penguin, 1977), 2.

3. Jaune Quick-to-See Smith, *The Submuloc Show/Columbus Wohs: A Visual Commentary on the Columbus Quincentennial from the Perspective of America's First People*, exhibition catalogue (Phoenix, AZ: Atlatl, 1992).

4. *Char-Koosta News*, Pablo, MT, June 5, 1992.

5. For a more complete description of the text collaged on the artwork, see Linda Muehlig's description of the painting in, Linda Muehlig, ed., *Masterworks of American Painting and Sculpture from the Smith College Museum of Art* (New York: Hudson Hills Press, 1999), 237–240.

6. The Dawes Act of 1887, also known as the General Allotment Act, divided reservation lands previously held in common into separate plots of land to be held by individuals; unallocated tribal lands were declared surplus and were opened for sale to whites. Only those able to document that they were one-half Indian were eligible for land and from that point forward, the federal government enforced eligibility requirements based on blood quantum, including enrollment as a member of a federally recognized tribe.

7. Sovereignty, identity, blood quantum, and tribal enrollment numbers also mark the artwork of many indigenous artists, including George Longfish, Da-ka-xeen Mehner, Edward Poitras, Hulleah Tsinhnahjinnie, Lyle Wilson, and Melanie Yazzie.

8. Alejandro Anreus, "A Conversation with Jaune Quick-to-See Smith," in *Subversions/Affirmations: Jaune Quick-to-See Smith*, exhibition catalogue (Jersey City, NJ: Jersey City Museum, 1996), 111.

9. Lucy Lippard, "Jaune Quick-to-See Smith's Public Art: Generosity with an Edge," in *Subversions/Affirmations: Jaune Quick-to-See Smith*, exhibition catalogue (Jersey City, NJ: Jersey City Museum, 1996), 79.

10. Jaune Quick-to-See Smith earned her Master's Degree in Art from the University of New Mexico in 1980.

11. Louis Owens, *Other Destinies: Understanding the American Indian Novel* (Norman, OK: University of Oklahoma Press, 1992), 94.

12. Joy Harjo, et al., *Reinventing the Enemy's Language: Contemporary Native Women's Writings of North America* (New York: W. W. Norton Company, 1997).

13. Lippard, 79.

14. Jaune Quick-to-See Smith, email message to the author, February 6, 2005. Jaune Quick-to-See Smith thinks that between 1991 and 1992, she created an edition of approximately twenty *Paper Dolls for the Post-Columbian World with Ensembles Contributed by the U.S. Government* in watercolor, pen, and pencil on photocopy paper, each measuring 17 × 11 inches.

15. It is worth noting that twice in the process of writing this essay, I have encountered the mistaken title of *Paper Dolls for a Post-Colonial World*.

16. Arthur Albert Smith was born in 1901 at the Métis colony at Pilcher Creek, Alberta, Canada. He was raised on the Flathead Reservation by his great-grandmother Nellie Quick-to-See Minesinger, his great-aunt Emma Magee, and his grandmother Mary Minesinger Miles.

17. Anreus, 108–09.

18. As Jaune Quick-to-See Smith lost her native language, she was estranged from her grandmother, Quick-to-See, who only spoke Salish.

19. Online Computer Library Center, Inc., http//www.worldcat.org (accessed on June 25, 2009). A search for "Barbie doll" produced hundreds of volumes written on the collectability and value of Barbie dolls, as well as those written by feminist theorists analyzing the impossibility of the false dreams expressed and invoked by the original Barbie doll.

20. Smith's knowledge of Western art history further powers her unique two-dimensional hybrid figures that mimic pictorial Indian stereotypes dating to the sixteenth century, and the beautifully rendered but inaccurate engravings of Theodore de Bry. See: Theodore de Bry's (1528–1598) illustrations of the "new world," first published in 1590 in a new edition of Thomas Hariot's *A Briefe and True Report of the New Found Land of Virginia*. De Bry's images were based on the botanical paintings of colonist John White and various descriptive texts. He visualized the people of the "new world" for Europeans, even though he had never seen them.

21. Jaune Quick-to-See Smith, e-mail message to the author, June 2, 2009. "The paper dolls are the story of my family and all of the families on Flathead."

22. Gerald Vizenor, quoted in Joseph Bruchac, *Survival This Way: Interviews with American Indian Poets* (Tucson: Sun Tracks and University of Arizona Press, 1987), 294. Vizenor has created a theory of Indian storytelling from the comic attitude of the trickster figure.

23. Anreus, 112.

24. Anreus, 113.

25. Gerald Vizenor, *Manifest Manners: Narratives on Post-Indian Survivance* (Lincoln: University of Nebraska Press, 1994), 83.

26. "Interview," in *Jaune Quick-to-See Smith: Made in America*, exhibition catalogue (Kansas City: UMKC-Belger Arts Center for Creative Studies, 2003), 6.

27. For more on the concept of the "trickster shift," see Allan J. Ryan, *The Trickster Shift: Humour and Irony in Contemporary Native Art* (Vancouver: UBC Press, 1999).

28. See Jaune Quick-to-See Smith's *Petroglyph Park Series*, 1986–89, and the *Chief Seattle Series*, 1989–91.

29. Anreus, 113.

30. Smith's strategy visually represents the effect of the Dawes Act of 1887, which preceded the forced removal of the Flathead people from their land by three years.

31. Jaune Quick-to-See Smith, e-mail message to the author, June 2, 2009.

32. Jaune Quick-to-See Smith, email message to the author, June 26, 2009. "By focusing on the Plenty Horses family (I used their name rather than Smith or any of the family French names because the name Plenty Horses immediately sets up a surreal polar twist with Barbie, the all-American Anglo girl. But truth is, this is my family story as well as the rest of my tribe.)."

33. Curriculum in Indian boarding schools was focused on vocational and domestic training. Students were in academic classrooms only a few hours each day until the 1930s. See Estelle Reel, *Uniform Course of Study for the Indian Schools of the United States* (Washington, DC: Government Printing Office, 1901).

34. Allan J. Ryan teaches his readers the seriousness of this action in his book, *The Trickster Shift: Humour and Irony in Contemporary Native Art* (Vancouver: UBC Press, 1999).

35. Ryan, 3.

36. Quoted in Ryan, 4.

37. Indian boarding schools were founded in the seventeenth century. During the 1880s, the federal government established the federal school system for Native American students.

38. A survey of only a few of these before and after photographs reveals the trickery of changing the skin color of the subjects in the "after photographs." See: Alfred L. Bush and Lee Clark Mitchell, *The Photograph and the American Indian* (Princeton: Princeton University Press, 1994), 90–91, 92–93, 104–105. For further analysis see: James Faris, *The Navajo and Photography: A Critical History of the Representation of an American People* (Albuquerque: University of New Mexico Press, 1996).

39. The Carlisle Indian School was established far from the reservations in Carlisle, Pennsylvania, in 1879. The founder of the school, Richard Henry Pratt, summarized the school's philosophy in one sentence. "Kill the Indian and save the man."

40. Jaune Quick-to-See Smith, email message to the author, June 26, 2009. This is a personal history that Smith details within her own family lineage. "If you consider that my father was born in 1901, his mother about 1870 and her mother about 1830. This was way before the crush of the Great Invasion and at that time there were no other tribes residing in Montana. We easily hunted and harvested on both sides of the Rockies, as well as traded with tribes up into Canada and all the way down River (the Columbia.) We were the most Eastern tribe of the largest trading chain in North America (lots of Salish speakers)."

41. Under the Garfield Treaty of 1891, the United States Military forced the Flathead people from their homelands in the Bitterroot Valley and moved them to a federally proscribed reservation.

42. Many authors have described this condition. See especially: Marsha Clift Bol, "Lakota Women's Artistic Strategies in Support of the Social System." *American Indian Culture and Research Journal* 9 (1985): 33–51. Richard White, *The Roots of Dependency: Subsistence, Environment, and Social Change among the Choctaws, Pawnees, and Navajos* (Lincoln: University of Nebraska, 1983).

43. From the eighteenth-century paintings by Benjamin West to the early twentieth-century photographs of Frank Hamilton Cushing, European Americans were pictured proudly with their collected Native American patrimony.

44. Simon Bronner argues that ethnological collections were powerful because they spoke the primary metaphor of "cultural evolution." Simon J. Bronner, ed., *Consuming Visions: Accumulation and Display of Goods in America 1880–1920* (New York: W. W. Norton & Company, 1989). Wanda Burch situates the power of early collecting between the two worlds of British aristocrats and Native North Americans.

Wanda Burch, "Sir William Johnson's Cabinet of Curiosities," *New York History,* July (1990): 261–82.

45. Jaune Quick-to-See Smith, e-mail message to the author, June 26, 2009. The dislocation from tribal lands and its effect on families reverberates throughout Jaune Quick-to-See Smith's personal history. She was raised by her father, Arthur Albert Smith, on the "Flathead Reservation in Montana, the Hupa in California, the Nisqually in Washington State and off reservation in many other places. As a horse-trader, he moved many times." Smith remembers counting perhaps 30 places she had lived by the time she was in high school and can no longer recount them all.

46. In addition to the bulwark of federal treaties and regulations, the United States government subsidized mission schools from 1810 to 1917.

47. *Jaune Quick-to-See Smith: Made in America,* 6.

48. Vizenor, *Manifest Manners,* 17.

49. Vizenor, *Manifest Manners,* 163.

50. Jaune Quick-to-See Smith, quoted in Anreus, 111.

51. Jaune Quick-to-See Smith, "Curator's Statement," in *The Submuloc Show/ Columbus Wohs: A Visual Commentary on the Columbus Quincentennial from the Perspective of America's First People,* exhibition catalogue (Phoenix, AZ: Atlatl, 1992), III.

2

———

Restating Indigenous Presence
in Eastern Dakota and Ho Chunk
(Winnebago) Portraits of the 1830s–1860s

Stephanie Pratt

This essay examines a small selection of painted and photographic portraits that represent Eastern Dakota and Winnebago male leaders from the southern Minnesota area in the middle years of the nineteenth century, between the 1830s and the 1860s. These portraits—two paintings and one photograph—were made in a period of some volatility and I will use them as a synecdoche for wider processes affecting cultural identity, its production and its reception, on the western frontier. In addition, by making reference to two distinctive portrait media (painting and photography) I wish to examine more precisely the interplay between the specific pictorial medium and the construction of identity by the sitter and the image taker. For it is in the details of this very particular encounter, what we might call the "moment" of portraiture, that we can witness the processes of identity construction in operation.[1]

In essence, the portrait artist in the inter-cultural situation represents the sitter as "other," the more so in that the sitter's personality and biography are not familiar, and especially when the portrait artist is from a dominant culture, the sitter from one whose culture and lifeways have been affected by the processes of settler-colonialism. In contrast, the sitter, even if s/he has not commissioned the portrait, normally wishes to project a sense of self that will be captured in the image.[2] Because this self-image is based on cultural considerations respecting personal and social identity, it is not something that a portrait artist from another culture will readily comprehend. The portrait itself, therefore, can be read as a contested site in which notions of "self" and "other" are held in tension.[3] It could also be viewed as one of the "arts of the

17

contact zone" as Mary Louise Pratt has framed them. As she states "[w]hile subordinate peoples do not usually control what emanates from the dominant culture, they do determine to varying extents what gets absorbed into their own and what it gets used for. Transculturation, like autoethnography, is a phenomenon of the contact zone."[4] In this light, portraits as construed in European terms via the painted or photographic image can be transformed in the indigenous cultures of North America into a means of self-expression and self-representation. Indeed, in a study of the portrait photographs taken of the Lakota (Oglala) chief Red Cloud, a case is made for the sitter's resistance and agency in creating a lasting visual legacy of himself as a memorial for his peoples. Photography presented Red Cloud with 'a powerful medium through which to communicate. Its reproducibility, accessibility and popularity made [it] additionally advantageous to this task."[5]

. The images under review in this essay were all made originally as part of wider projects to record and document Native Americans. Whether painted or photographed, the ideological presumptions that underpinned them remained constant: that the American Indians' way of life on the land was doomed, that their traditional cultures would not be sustained in the modern United States, and that as a distinctive social fraction they would inevitably decline and in all probability vanish, whether through a process of assimilation or because of the ravages of disease or alcoholism. The business of taking likenesses was predicated on the assumption that these peoples were culturally distinctive and as such the proper subject of disinterested ethnographic enquiry. The painted and photographed portraits that were produced in this spirit made a virtue of their accurate record and deployed a limited repertoire of stock poses to allow cross-cultural comparisons to be made.

The making and collecting of such physiognomic data banks has been analyzed to good effect with respect to the correspondence between colonial and anthropological enquiry in the nineteenth century. Important essays on the photographic imaging of Native Americans by Buerger, Dippie, Fleming, and others have helped indicate the relationship between the artist/photographer and the wider epistemic horizon within which this work was produced, circulated, and received.[6] This chapter contributes to that literature by speculating about the sitter's employment of material artefacts as items of symbolic capital whose specific meaning and resonance, by virtue of their highly localized integration into tribal knowledge systems, escapes the domineering gaze of the artist/photographer. Depending on circumstances, these items may therefore be understood either as implicitly oppositional— "untranslatable" goods that by definition escape the desired transparency, or "readability" of the constructed image—or as objects deliberately chosen by the sitter to frustrate the process of reading the image from "outside" and therefore acting as a form of cultural resistance.[7]

The three subjects of these portraits were all distinguished individuals, occupying significant positions in their communities. The first sitter to be discussed is an Eastern Dakota chieftain, "Wa-nah-de-tunk-ah" or Wamditanka, Big Eagle, also known as Black Dog, who was painted by George Catlin in 1835[8] (Figure 4, photo gallery). The second sitter is a Winnebago/Ho Chunk man, Wa-chee-háhs-ka, or the Boxer, a distinguished war leader, who was also painted by Catlin in 1835[9] (Figure 5, photo gallery). The third sitter under review here is the first-mentioned sitter's grandson, also a chieftain, Wamditanka, also called Big Eagle or Jerome Big Eagle, who was photographed twice between 1860 and 1864[10] (Figure 6, photo gallery). I will discuss how these portraits can be related to each other, how they have been interpreted in the past according to Western art historical paradigms and, crucially, how the objects of material culture depicted in them offer alternative avenues toward a more balanced and reciprocal interpretation of the portraits' meanings.

As a starting point, we need to remember how any portrait image is saturated with pre-given (Western) expectations of its function and purpose. Many of the images of American Indian men found in the historical records, from the earliest era of contact to the mid-nineteenth century, conform to standard European portrait formats, whether seated or standing. They use paraphernalia and other items of accoutrement to signify the status of the sitter and often contain within them references to existing iconographies of the leader or ruler, as codified in the Western tradition. Arguably, this convention begins with John White's watercolors of Algonquians in Virginia (now part of coastal North Carolina) in the 1590s, always accepting that in the England of White's time portraiture was in its infancy.[11] After White, and especially in the eighteenth and early nineteenth centuries, a number of English artists took the likenesses of Indian visitors to London, especially those who arrived on diplomatic embassies to secure alliances with the British against the French and, latterly, the Americans.[12] Between 1822 and 1830, so immediately before Catlin's project got underway, the American artist Charles Bird King made the first large portrait collection of Native Americans when he painted successive Indian delegates to the federal government in Washington for the Superintendent of Indian Affairs, Thomas L. McKenney. Portraiture was now a ubiquitous art form, recording not merely the most powerful members of society but also the rising middle class as well. Its procedures were therefore a commonplace of Western experience and both artist and sitter worked with a shared understanding of what is involved in taking a likeness. Its extension to Indian subjects was to be expected, but of course the relationship between artist and sitter could not operate with the same kind of shared understanding and familiarity.

The extension of the practice to other communities, for whom illusionistic portraiture was not a recognized activity, necessarily resulted in another negotiation entirely. The portrait image always mediates identity, but within the Western tradition both parties (artist and sitter) collude in its results. When a portrait artist takes the likeness of an individual differentially subjected to his gaze, the possibility of such a mutually accepted collusion is remote; in such circumstances the sitter's attempt to control the image is radically restricted. In nineteenth-century America it was portrait painters and photographers, trained in the Western tradition, who recorded the likenesses of Native peoples. The images that result show us what they proposed as an adequate representation of these individuals. But if we use material culture to widen the aperture of perception, it may be that we can discern something that lies mute within these images; the residue of what the sitter brought to the occasion. These items, irrespective of the Western artist's limited comprehension of them, may register concepts of identity that largely escape the limitations of the western portrait tradition.

For the Western viewer, in other words, the reception of Catlin's chieftains and war leaders is guided by the ways that the artist can suggest status in the figures' poses and regalia, their "crown" or assortment of hair adornments and feathers, peace medals, pipes or pipe-tomahawks, Hudson Bay blankets, or other gifts and honors. But these same objects would have signified very different things to a Native American viewer. Items of material culture had a significance and meaning that was as much discursive as it was ceremonial or ornamental. As Max Carocci has recently revealed, most items of Native American material culture existing in museum collections today cannot be relegated to the category of "things" or mere "objects" in the Western sense of those terms. Instead they must be viewed as forms of living history, recast as ancestor-receptacles and denoted as "thing-beings" escaping all forms of Western description and designation.[13] When Western scholars refer to such items as a decorated buckskin shirt or crafted weaponry as potential "thing-beings," they are in effect pointing to a recently revised anthropological approach to the interpretation of such "things." Such an approach stems largely from the work of Marilyn Strathern and others in the field of social anthropology who proposed more relational models in place of the then established anthropological conception of material culture.[14] In a similar vein, important studies have appeared, taking on the challenge of re-imagining the worldview of "others" whose understandings often differed from the Cartesian and similar classical models developed in the West for the understanding of human reality and the creation of meaning. In the early work of A. Irving Hallowell on Ojibwa ontologies, for example, and more recently in Robert Wallis's account of rock art, a model of interpretation has been proposed which takes account of the "personhood" of particular objects or material substances.[15] In Hallowell's analysis of Ojibwa worldviews he revealed that the terminology used for both "humans" and "other than humans" refers to the same ontological status.

This helps to disclose the continuity of "substance" of both these aspects of reality (the human and other than human) for Ojibwa religious practitioners and thus to disclose an "object's" explanatory powers. Subsequent work has promoted the idea of "seeing subjects and objects as mutually constitutive." a development that marks an important shift in the discipline.[16]

In the case of the three portraits I have selected, it is likely that many of the indigenously construed items on display in them carried specific connotations for the Native American person (whether sitter or viewer) and that their interpretation of them extended to other matters besides social status, hierarchy, or singular command. In the Western tradition we expect to decode accoutrements as personal possessions or markers of individual identity; in the Native American tradition they speak more of communal, ancestral, and transcendent signification.

As a first step, then, it makes sense to look for characteristics in Indian self-presentation. If we begin with the earlier portrait images by George Catlin of *Big Eagle* and *The Boxer* in 1835, and link these to the later photograph of *Jerome Big Eagle* in 1864, seeing these images in relation to one another, a better understanding can be gained of the ways in which the inclusion of material culture items work as a signifying practice. And, in paying attention to the kinds of objects each sitter displays on and about their person, we can begin to comprehend how objects can represent relationships outside and beyond the frame of the portrait.

The figure of Big Eagle or Black Dog as depicted by Catlin in 1835 shows a particular individual whose name is titular rather than personal (Figure 4, photo gallery). Wamditanka was the name given to the leader of the Oanoska (meaning "Great Avenue") band of Dakota, and their village, called Black Dog, which sat traditionally along the St. Peter River.[17] Dr. William Keating, of the Stephen H. Long Expedition team of 1823, met many of the Eastern Dakota leaders in this region and briefly mentioned a man called "War Eagle," who is probably the same person Catlin portrayed twelve years later. In Catlin's portrait, Big Eagle wears his hair loose about his shoulders and has trimmed his fringe very short across his forehead, a style worn by the Dakota men as noted by Samuel Pond in the published narrative of his time in southern Minnesota in 1834.[18] Big Eagle wears a single feather in his hair, probably an eagle feather, and a large roach made of the warning hair of a porcupine, painted red, and pasted into a deer's tail hairs. This was widely used by men as head ornamentation across the entire Western Indian country at this time. There are also trade items displayed in Catlin's portrait of Big Eagle, particularly the white wool Hudson's Bay Company blanket he has draped around his shoulders and the peace medal, or American Fur Company medal that he wears around his neck. Probably the most important item he shows on his person in Native American terms is the pipe or *chanumpa* that he holds across his left arm. The leader's pipe would have been given to him from clan elders with a "bundle" containing the spiritual power of the office that the leader held for his people. The pipe stem Catlin shows here looks to be of the flattened design associated with the Dakota and decorated with colors and patterns in quillwork and beading. The pipe bowl would have been carved from the soft red rock found at a special quarry, now called Pipestone. This mineral was (and still is) believed to be a living substance and to signify the physical body of Mother Earth. One's *chanumpa* is tied to one's own existence and health and has healing powers. However, in this portrait the sitter is shown holding the pipe in an uncharacteristic manner with the pipe bowl pointing upward toward his right shoulder.[19] This would never have occurred in actual Dakota ceremonial as the pipe bowl remained pointed downward in the smoking position and was always passed to another via holding the bowl and stem in a horizontal manner. To do otherwise would be and still is seen as disrespectful. Its visualization in this manner points to the intervention by Catlin who may have wished to demonstrate that his sitter was an important figure, being a chief and "pipe-carrier." As his normal procedure was to paint only the sitter's face and head in situ and to complete the portrait later, adding details of clothing, weapons, and material goods, such an error is readily comprehensible.

We can contrast this image with the second of Catlin's portraits, that of the Winnebago/Ho Chunk man called The Boxer, who Catlin revealed died tragically of smallpox just after this portrait was made (Figure 5, photo gallery). The Boxer or 'Wah-chee-háhs-ka' (Man who puts all Out-of-Doors), appears to be a much younger man than Big Eagle. He wears his long dark hair braided in two plaits and hanging down his back, in keeping with his youth. His hair looks to be cut short and raised up stiffly over the crown of his head, or there may be a small roach attached there. His hair contains many more eagle feathers than that of Big Eagle and this suggests more frequent wartime activities and successful coups earned against enemies. On each wrist he has attached the carcass of a dead rattlesnake and in his right hand he grasps the end of a war club made from a gunstock, probably acquired through trading outlets in the area. All of these factors point to his success as a warrior and his style of dress, with his upper body left bare and wearing a breechcloth and knee-length leggings with moccasins on his feet, shows how he might have appeared in battle. The pipe bowl and stem that Catlin has given him are shown in the correct orientation for smoking, but they are unusual and very unlike the typical Dakota or Plains style. However, there are examples of curved stemmed pipes like these made by the Pottawatomi peoples of the Great Lakes region. This is not unexpected as the Winnebago or Ho Chunk people originally lived and hunted in western Wisconsin near the Southern Great Lakes region and may have traded for this item within the region.

Finally, in the 1864 photographic portrait of Wamditanka or Jerome Big Eagle, we see the grandson of Wamditanka, Big Eagle (Figure 6, photo gallery). His own father, Gray Iron, had given him the title of Wamditanka, and so the young man shown here took up an office that had existed many generations before him. However, in the photograph we do not see similar sorts of adornment or trade items displayed on his figure, as they were when his grandfather sat to Catlin in 1835. Jerome Big Eagle appears to be closer in dress and ornamentation to the Winnebago/Ho Chunk man, Wa-chee-háhs-ka, than to his grandfather and this reflects the recent mingling and reciprocity between Dakota and Winnebago cultures in the intervening thirty years.[20]

In terms of Native American sensibilities and understandings, a chieftain made his mark as a leader through his deeds and the qualities of his leadership and these are elements that a portrait created by a western artist cannot completely depict. However, if we are able to conceive of such items as a trade blanket, pipe-tomahawk, war club, or a peace medal as being "intermedial" in their representations (i.e., creating meaningful information across a number of modes/forms of expression), is it not also possible to read out of such objects meanings that would "make sense" from a Native American perspective? I would like to suggest a procedure that could reveal how such imagery would have been conceived by its Native American audiences.

A small measure of understanding of this process can be garnered from the few written records detailing Native sitters' reactions to European-style portrait making. There are certain well-known examples of this that are frequently mentioned in the literature, such as the statement by the Cherokee warrior, Cunneshote, who had his portrait painted in London in 1762, stating that "he was glad that his family would have something to remember him by when he had gone off to fight the French."[21] Cunneshote presumably knew that the oil painting was destined for a British collection, but he must also have been informed that portable versions of his portrait would be produced in the form of the mezzotint print made by James MacArdell. His comment suggests that he understood the main function of European portrait making to be a means of memorializing those of distinction. Cunneshote's understanding of the Western portrait's purpose may be contrasted with George Catlin's story of the two Lakota men, Little Bear and The Dog, who initially interpreted a three-quarter profile portrait of Little Bear as status-driven, the "missing" part of the face omitted because it was unimportant, and then credited the portrait with powers of prediction after the same side of Little Bear's face was destroyed by a gun blast.[22]

In this context, the life and career of Zacharie Vincent (Telari-o-lin) is instructive. Vincent (1815–96) was a Huron (Wendat) chief from Jeune Lorette (now known as Wendake), a settlement close to Quebec. Originally founded in 1697, by the nineteenth century Jeune Lorette had become a popular tourist destination; visitors bought Indian goods and numerous depictions were made of the village and its inhabitants. The community was predominantly Catholic by religion and French-speaking; intermarriage with whites meant that by the 1800s the population was effectively métis. Vincent, however, was encouraged by his father to preserve the language and the traditional culture of his people and was wont to refer to himself as "The Last of the Hurons."[23] Antoine Plamondon painted his portrait and exhibited it in May 1838 as *Le dernier des Hurons* at the Société Littéraire et Historique de Québec, where it was awarded a first-class medal and bought by Lord Durham. Plamondon's portrait shows Vincent in traditional costume gazing in contemplation at the sky but, for all its intrinsic interest in the circumstances of the 1830s, the picture is relatively generalizing in its treatment of native dress.[24] When compared to the self-portraits Vincent painted, however, Plamondon's approach is thrown into relief. Vincent was given some lessons by Plamondon and he left behind a body of some 600 drawings and paintings, showing the landscape around Lorette, the activities of his community, and also a number of self-portraits, usually inscribed "Telari-o-lin." In these latter especially we can see how he might position himself as a synecdoche of his culture, presenting Huron identity to a Western audience in a Western idiom. What is especially notable about these self-portraits is the emphasis given to items of material culture, irrespective of whether these are "authentically" Huron or constitute the hybridical culture of the mid-nineteenth century. Painted by someone who knew exactly what they signified, these weapons, medals, wampum, and ornamental objects are depicted with a certainty and conviction that no Western artist could have attained[25] (Figure 7, photo gallery).

The significance of material culture for Indian self-representation can be further underlined if Western modes of image-making are compared with Indian pictographic art at this time. The ideograms characteristic of this art tended to show the individual frontally and only animals in profile. In graphic depictions found on war shirts (e.g., biographic art) or in winter counts (an example of "collective memorialization"), there is an emphasis on the action being depicted rather than the individual's unique facial attributes. Both styles of art could be thought of as "portraiture" in that the pictograph makes narrative reference to an individual and their history.[26] In the pictograph, historical significance and narrative emphasis might be given to what a Euro- or Anglo-viewer would have thought were small or even insignificant changes in design, such as the placing of a heavy coat across the body of a figure (meaning either wintertime or men going on the warpath) or the cutting off of the head of one figure in a chain of very similar figures (to suggest the execution of a captive).[27] With reference to the photograph of Wamditanka, Jerome Big Eagle, taken in Washington DC, it is significant that in Plains cultural imagery pipes were often shown at the owner's waist. For example, in a Mandan buffalo-skin robe, documenting the exploits of Mato-tope (or Four Bears) he is shown upright with his pipe painted horizontally across his waist, unsupported by his hands[28] (Figure 8, photo gallery). This placement accords with the horizontal handing over of the pipe in ceremonies.

With these sorts of differences in mind, it seems more in keeping that the men who Catlin painted would have viewed their image in the portrait in ways similar to their own traditional concepts of representation. If so, the facial features would have been read as less significant than the disposition of the body and the variety and quality of clothing and accoutrement.

Looking between the two portraits of Big Eagle and his grandson, we see several similarities in their representation. Both men are shown looking forward solemnly and displaying the signs of Native American manly distinction in wearing coup feathers in their hair. For the Western viewer, these are the most significant markers of their Indian identity. But they also wear strings of beads and/or a peace or trade medal worn around the neck, items of trade and negotiation. In addition, as we have seen, both men also carry a significant item in their hands: Big Eagle has a pipe and his grandson holds a gunstock club. All of these possessions reveal the workings of the zone of contact on the frontier, where trade and exchange fostered the growth of intercultural understandings. The complex web of contact, trade and negotiation of the "middle ground" produced a dynamic environment that fostered in its participants self-fashioned hybridical identities. Seen in this light, the material items depicted in the portraits are better understood as discursive signs or even living, real presences as were the sitters themselves, whose meaning can only be understood in the context of a particular moment in frontier history. Furthermore, in drawing attention to material culture a more indigenously-orientated and reciprocal set of readings can be undertaken.

In Jerome Big Eagle's portrait, he sits straddled across a stool or seat and is facing us frontally. This kind of self-representation would tie in more closely with a pictographic type of portrait rendition, especially in terms of the way a warrior would show himself with the emphasis on his weapon or his war "deeds" not particularly on his own physiognomy. He wears six eagle feathers in his hair specifically to signify the six Ojibwa warriors he defeated or touched (counted coup upon) in battle. The feathers worn in the hair at the top of the head were given for such wartime or horse-stealing accomplishments, which each individual could recount and relive in that way. The feathers not only had visual power to encode these actions but could also be used again and again to invoke *wakan* or the spirit of the bird.

Crucially, he holds in his arms a gunstock club telling of his status as a warrior (and thus not so much a civil chieftain), but the blade of his club has been removed somewhat uncharacteristically given the number of images of Indian leaders which include it, and in this presentation it becomes less of a violent weapon. The removal of the blade may be very important if we consider its significance from a pictographic or ideographic basis. Although Jerome Big Eagle is showing himself as a warrior in this portrait (he had just been involved in the 1862 Dakota uprising when the portrait was taken), in Native American terms the lack of this essential part of his weaponry effectively pacifies the figure shown here. He is a warrior without purpose for he cannot be an effective soldier with his war club laid to the side, much as the pipe held by his grandfather earlier had signified a peaceful stance (even if held inaccurately in ceremonial terms) from one of the signatories to the Treaty at Prairie du Chien in 1825. Read in another way, however, one might see resistance in Jerome Big Eagle's war club in that the removal of the metal blade might also mean rejection of a European-American produced trade item. Hence, it might mean Jerome's insistence on a more Native-based understanding of warfare and indigenous signification systems.

Jerome Big Eagle's life, however, spanned some of the most disruptive events of nineteenth-century Dakota history. Merely listing some of these changes gives some idea of the volatile years leading up to and including his time as leader. The long-held fur and other trade relationships were evolving from Native-based into more capitalistic models in the 1840s. Protestant missions were widely established amongst the main Dakota settlements from the 1840s and 50s.[29] Several major treaties, giving land rights to further expansion of white settlement, and to which Big Eagle was a signatory, had been enacted in the 1850s. Finally, out and out warfare between Dakota and white settlers erupted in 1862.

Jerome Big Eagle had been one of several leaders in 1862 forced to make a choice between continuing with the acculturation processes already set in motion through treaty and Christianization, or in walking the traditional path of the chief in being selfless and completely beholden to the majority wishes of his people. Jerome Big Eagle made the choice to go on the warpath at this time, but also withheld himself from attacking and killing anyone he would have seen as helpless or a non-combatant, such as women and children or those he considered his friends amongst the white settlers or traders. At the trial following the 1862 events, he was defended by witnesses who supported his claims; he thus escaped execution and was only given a three-year prison term. (Nearly 400 men were originally condemned to death but President Lincoln commuted most of these to imprisonment. A total of 38 men were hanged at Mankato in Minnesota after a show trial took place in 1862.) Big Eagle's portrait photograph of 1864 thus reveals someone poised between the rapidly advancing acculturation of his small band and of the Dakota themselves, and a return to the old ways if it were possible. However, he maintains his status and agency through measuring himself against Native

American expectations and systems of representation, albeit that these were nearly annihilated at this time.

In conclusion, what these examples show is the possibility of sitters to portrait-takers maintaining some sort of agency over their representation, even within the severely restricted space of articulation left open to them. This is not to discount the force of the western image-making process, its cultural pressure and the discursive matrix from which it emanated. But it is to suggest that those subjected to the artist or photographer's gaze were entirely capable of working against the grain of the ideology that supported it, resisting seamless incorporation into the confected world of the Western image. From a methodological point of view, this possibility is necessarily elusive by virtue of the fact that the objects whose material presence disrupts the authority of the Western portrait speak without texts, constituting a mode of knowledge that is particular to the performance of a culture. Working at this remove and lacking other sorts of testimony, analysis can rarely be conclusive. However, there is sufficient evidence pertaining to material culture items in other circumstances to afford some reassurance that these sorts of speculative enquiries may at the very least help to reposition our thinking about the portraiture of Native American peoples. More generally, one might view this kind of enquiry as further developing the kinds of analyses associated with subaltern studies, replacing the binary of oppressor/oppressed with a more nuanced understanding of the processes of mediation between cultures and within cultural products, such that the recognition of "agency" in the object of the gaze moves the sitter from a situation of complete passivity to one of interlocutionary presence.

Afterword: On his release from prison in 1865, Jerome Big Eagle became the second husband of my great-great Grandmother, Whirlwind in the House, or Mary Snow Deloria (Agar).

Notes

1. This essay draws on the general literature concerning photography and its use as an anthropological and ethnographic tool for recording Native American lifeways and physiognomies: see, for instance, Paula Fleming and Judith Luskey, *The North American Indian in Early Photographs* (Oxford: Phaidon, 1988) and Paula Fleming and Judith Luskey, *Shadow Catchers* (London: Calman & King, 1993). For an indigenous response to nineteenth- and early twentieth-century photography, see "Introduction" by Lucy R. Lippard, ed., *Partial Recall: Photographs of Native North Americans* (New York: New Press, 1992), 13–45; and Gerald Vizenor, *Fugitive Poses: Native American Indian Scenes of Absence and Presence* (Lincoln, NE and London: University of Nebraska Press, 1998). General studies of the photographic representation

of Native Americans include: Jane Alison, ed., *Native Nations: Journeys in American Photography* (London: Barbican Art Gallery/ Booth-Clibborn Editions, 1998) and Alfred L. Bush and Lee Clark Mitchell, *The Photograph and the American Indian* (Princeton: Princeton University Press, 1994).

2. See the discussion of the inherent subjectivity in urban poor sitters to portrait photography in Eric Homberger, "J. P. Morgan's Nose: Photographer and Subject in American Portrait Photography" in *The Portrait in Photography,* ed. Graham Clarke (London: Reaktion Books 1992), 115–131.

3. Ideas of "transculturation" and concepts of shared meanings within a "middle ground" of intercultural contact were first iterated respectively by Mary Louise Pratt, *Imperial Eyes. Travel Writing and Transculturation* (London: Routledge, 1992); and Richard White, *The Middle Ground, Indians, Empires, and Republics in the Great Lakes Region, 1650–1815* (Cambridge and London: Cambridge University Press, 1991).

4. Mary Louise Pratt, "Arts of the contact zone" *Profession* 91 (New York: MLA, 1991), 33–40.

5. Frank H. Goodyear III, *Red Cloud. Photographs of a Lakota Chief* (Lincoln, NE and London: University of Nebraska Press, 2003), 4–5.

6. There are a number of recent studies which examine the conjunction of photography, anthropological visual culture, and the imaging of Americans Indians. The most important of these are: Melissa Banata and Curtis Hinsley, *From Site to Sight: Anthropology, Photography and the Power of Imagery* (Cambridge, MA: Peabody Museum Press, 1986); Brian Street "Representing the other: the North American Indian"; Brian W. Dippie "Of 'peculiar carvings and architectural devices': photographic ethnohistory and the Haida Indians"; and Margaret B. Blackman."George Hunt, Kwakiutl photographer" all in *Anthropology and Photography, 1860–1920*, edited by Elizabeth Edwards (New Haven: Yale University Press in association with the Royal Anthropological Institute, London, 1992); Janet E. Buerger, "Ultima Thule: American Myth, Frontier, and the Artist-Priest in Early American Photography," *American Art*, v. 6, n. 1, Winter (1992), 82–103; Cory Willmott "The Lens of Science: Anthropometric Photography and the Chippewa, 1890–1920," *Visual Anthropology*, 18 (2005), 309–337. And, for a comparative example, see Jane Lydon *Eye Contact: Photographing Indigenous Australians* (Durham, NC: Duke University Press, 2005).

7. It is of course well-known that many portrait photographers such as Edward Curtis carried with them items of indigenous production with which to accoutre the subjects of their photographs in many instances. Notwithstanding this very constructed view of the image of American Indians, it is still the case that indigenous meanings might be generated from items held by a sitter, even if the item in question is not authentically linked to the sitter's originating culture or society. The author wishes to thank Dr. Max Carocci for drawing attention to this problematic about the material culture items seen in many photographs of Native American sitters.

8. The artist George Catlin travelled widely in the Trans-Mississippi West between the years 1830–1838 depicting American Indians in painted portraits from many western areas, but principally the Plains. His use of a hyphenated and phonetically constructed approximation of an individual's name has been used here alongside the modernized Dakota term "Wamditanka."

9. Catlin's 1835 trip to Fort Snelling on the St Peter's and Minnesota Rivers, near present day Minneapolis, is detailed in his *Letters and Notes on the Customs and Conditions of North American Indians*, vol. II, 2nd ed. 1844 (New York: Dover Publications, reprinted 1973), Letter no. 50, 129–34; his trip to Prairie du Chien where he painted the portrait of Wah-chee-hahs-ka is detailed in Letter no. 52, 141–47.

10. We know from a framed photograph of Wamditanka (Jerome Big Eagle) that his photograph was also taken in the Studio of Robert W. Addis, who set up his photographic practice on 308 Pennsylvania Ave, Washington, DC from 1860–64; see Peter E. Palmquist and Thomas R. Kailbourn, *Pioneer Photographers of the Far West, A Biographical Dictionary, 1840–1865* (Palo Alto: Stanford University Press, 2000), 79.

11. For an examination of John White's watercolors of Algonquian Indians in coastal North Carolina, see Stephanie Pratt, "Truth and Artifice in the Visualization of Native Peoples: From the Time of John White to the Beginning of the 18th Century," in *European Visions: American Voices*, ed., Kim Sloan (London: British Museum Press, British Museum Research Publication 172, 2009), 33–40.

12. For a discussion of such images and their contexts, see Stephanie Pratt, "'The Four Indian Kings'" and "Joseph Brant (Thayendanegea)" in *Between Worlds: Voyagers to Britain, 1700–1850*, ed., Jocelyn Hackforth-Jones (National Portrait Gallery publications, London, 2007), 22–35 and 57–67; and Alden T. Vaughan, *Transatlantic Encounters, American Indians in Britain, 1500–1776* (Cambridge and New York; Cambridge University Press, 2006).

13. Max Carocci, "Introduction" to *Warriors of the Plains* (London: British Museum Press, 2012), 7.

14. See, Marilyn Strathern, *The Gender of the Gift* (Berkeley: University of California Press, 1988) and subsequent discussion of the problematic of approaching "things" as having aspects of "personhood" in Amiria Henare, Martin Holbraad and Sari Wastell (eds.) "Introduction" to *Thinking Through Things, Theorising Artefacts Ethnographically* (London and New York: Routledge, 2007) 1–31.

15. A. Irving Hallowell, "Ojibwa Ontology, Behavior and World-view," in Stanley Diamond (ed.) *Culture in History* (New York: Columbia University Press, 1960) 19–52); and Robert J. Wallis "Re-enchanting Rock Art Landscapes: Animic Ontologies, Nonhuman Agency and Rhizomic Personhood," in *Time and Mind: The Journal of Archaeology, Consciousness and Culture* (London: Berg), 2 (1), March 2009: 47–70.

16. This more radical approach to the study of things is proposed in Henare, et al. (eds.) "Introduction" to *Thinking Through Things* (London and New York: Routledge), 2–4.

17. As listed by William H. Keating in *Narrative of an Expedition to the Source of the St. Peter's River, . . . performed in the year 1823* (from notes compiled by the author) (Philadelphia, PA: 1824), 385, accessed on 15 August, 2011 from the Library of Congress Web site at http://memory.loc.gov/cgi-bin/query/r?ammem/lhbum:@field(DOCID+@lit(lhbum1607adiv18).

18. Samuel W Pond, *The Dakota or Sioux in Minnesota as They Were in 1834* (Minneapolis/St. Paul: Minnesota Historical Society Press, 1908), 36

19. The author is indebted to Sharon Many Fingers Venne (Cree) for sharing this observation about the uncharacteristic mode of holding his pipe in Wamditanka's portrait.

20. Both the Dakota and the Winnebago speak a similar Siouan language. It is known that they traveled together and made coalescent groupings during the mid-nineteenth century, and according to some contemporary commentators they were at one time the same people.

21. See Stephanie Pratt, " 'Reynolds' 'King of the Cherokees' and other mistaken Identities in the Portraiture of Native American Delegations, 1710–1762," *Oxford Art Journal*, 21, 2 (1998), 133–150.

22. The story of The Dog (Shonka) and Little Bear is told in Letter no. 55 (from Red Pipestone Quarry, Coteau des Prairies) in George Catlin *Letters and Notes on the Customs and Conditions of North American Indians*, vol. II, London, 2nd ed., 1844 (New York: Dover Publications, reprinted 1973), 188–194.

23. Traditionally it was assumed that Telari-o-lin meant "pure blood," but this assumption has been undermined by recent research. See Louise Vigneault "Zacharie Vincent: Dernier Huron et premier artiste autochtone de tradition occidentale," *Mens*, 6, 2 (printemps 2006), 239–261.

24. Zacharie Vincent, *Self-Portrait*, mid-nineteenth century, oil on canvas, Musée de La Civilisation, dépôt du Séminaire de Québec (Accession number: 1991.102). For the meaning of this picture see François-Marc Gagnon and Yves Lacasse "Antione Plamondon *Le dernier des Hurons* (1838)," *Journal of Canadian Art History*, 12, part.1 (1989), 68–79.

25. For information on Vincent see Vigneault "Zacharie Vincent: Dernier Huron et premier artiste autochtone de tradition occidentale," 239–261.

26. For an explanation of the ways that pictographic art and culture can be viewed as autobiography, see Hertha Dawn Wong, *Sending My Heart Back Across the Years, Tradition and Innovation in Native American Autobiography* (New York and Oxford: Oxford University Press, 1992).

27. Max Carocci, "Visualising Gender Variability in Plains Indian Pictographic Art," *American Indian Culture and Research Journal*, 33, 1 (2009), 4.

28. Colin Taylor *Buckskin and Buffalo: The Artistry of the Plains Indians* (London: Salamander Books, 1998), 12.

29. For reference to the impact of the fur trade in the upper Minnesota and Missouri River areas, see Mary K. Whelan, "Dakota Indian Economics and the Nineteenth-Century Fur Trade," *Ethnohistory*, 40, 2 (Spring, 1993), 246–276; and for locations of the Protestant Missions (from 1847 until 1862) in this area, see the Bloomington, MN city Web pages at http://www.ci.bloomington.mn.us/main_top/2_facilities/rec_facility/pond/signs/missions/missions.htm accessed on 15, August, 2011.

II

DISMEMBERMENT AND DISPLAY

3

Plaster-Cast Indians at the National Museum

Jacqueline Fear-Segal

The artist James Luna (Luiseño) uses his own body as a performative instrument in order to deconstruct historically entrenched stereotypes of Native identity and so counterbalance the objectification of the Native body in museum displays. In one of his best known works, "Artefact Piece," 1987, he delivered a set of complicated messages to his viewer. Lying face-up, near-naked, and motionless in a glass exhibition case, Luna invited inspection of the apparently-dead Indian body at close range. He carefully positioned labels around himself, supplying his name and other vital statistics. In two nearby cases, he laid out for inspection his ceremonial items and regalia, as well as personal documents and memorabilia—degree, college mementos, divorce papers, photographs, record albums, and cassettes. As Luna explained in an interview, these told "a story about a man who was in college in the 1960s," but, as he points out, "this man happened to be native, and that was the twist on it."[1] Luna was inviting museum visitors to reflect on the clash of meanings between the separate displays when read alongside each other. Visitors to the Museum of Man, in San Diego, where Luna lay on display for several days alongside Kumeyaay exhibits, were to witness another "twist."[2] Periodically, Luna sat up, climbed out of his case, and walked around, using his body as a vigorous, kinetic revelation that the Indian body on display was neither dead, nor restricted to his glass case. With a humorous nod to the notion of a ghost rising from the tomb, Luna caught some visitors in the intrusive act of scrutinizing his body. Others he surprised in the process of visually rifling his personal possessions; at this moment, his clothing and personal papers suddenly and dramatically changed registers, shifting from being archival artifacts to become the personal and documentary accessories of a living individual. As Luna moved about the room/museum, visitors were

33

forced to acknowledge his presence as a living, breathing Native American. His eventual return to his glass case, however, made oblique reference to the continuing power of the forces that had placed him there; its transparent sides ensuring that his exposed body continued to be on display.

In "Artifact Piece," Luna stages an ironic contemporary reminder of how Native American bodies have been objectified in museums and the ways in which apparently "innocent" visitors are implicated in furthering this process. His prime purpose in this performative artwork is to educate his white audience. As Luna puts it, "I guess the statement is that I'm not up here to entertain, though I can be damn entertaining. I'm here to teach you."[3] Just as nineteenth-century museum displays incorporating Indian bodies had been set up to instruct the viewing public in systems of racial hierarchy, so too in the twentieth century, Luna used the display of his body as a means to re-educate his audience by disrupting and subverting these same quasi-scientific systems of hierarchy, while alerting the viewers to their own complicity in perpetuating such systems.

Through shock-tactics, humor, and allusion to well-known traditions of museum display, Luna invited his contemporary viewers to grapple with an array of complex historical and racial issues associated with public exhibition of the Indian body. The impact of his "Artifact Piece" depends on a wide range of historical and cultural references, all deeply rooted in the American past. First, by positioning himself as an Indian "specimen" for examination, Luna draws attention to the asymmetries of power implicit in museum displays. These derive from the colonizing relationships that underpin them, which deprive the subject of both humanity and legitimate the appropriation of bodies and body parts for visual consumption and display. Second, Luna's labels, and the secondary cases full of the accoutrements of his life, provide an ironic commentary on the use of the colonizers' specialized systems of knowledge to classify and categorize indigenous bodies and artifacts of material culture. Third, his display demonstrates how Indian bodies and artifacts are utilized as "authentic" evidence to construct Indian difference and racial hierarchy, and present them to the public as "scientific" and therefore objective realities.

The triangulation of these issues provides the backdrop for this study of the Smithsonian's National Museum in the late nineteenth century, where displays of indigenous bodies, body components, and material cultures were organized for public viewing. Focus will on be a specific component of the Smithsonian's Ethnology exhibit—a set of Native American plaster heads, commissioned and then created using life-masks taken from sixty-four Plains Indians being held prisoner in Fort Marion, Florida. Analysis of the history and public display of these physical replicas of leaders and warriors from recently conquered tribes reveals some of the complex processes by which Indian difference and inferiority was constructed and exhibited in this National Museum to bolster and legitimate the narrative of national expansion.

In his classic work on *Objects and Others,* George Stocking reminds us that "there are relations implicit in the constitution of a museum, which may be defined as relations of 'power'":

> Whatever the contingencies of their specific histories, the three-dimensional objects thrown in the way of museum observers from out of the past are not placed there by historical accident. Their placement in museums, their problematic character, and indeed, their "otherness," are the outcome of large scale historical processes.[4]

In the final quarter of the nineteenth century, these "large scale historical processes," which enabled both the creation of the Smithsonian and the amassing of its vast collections, were underpinned by the same forces of economic development and geographical expansion that were transforming and empowering the United States (and Europe), and simultaneously promoting assumptions about the superiority of Western technology, culture, and peoples. The Smithsonian's displays appeared to be presented and organized according to the "orderly and authoritative principles of science—principles conceived of as separate from power and politics."[5] But, as Sharon Macdonald makes plain in *The Politics of Display,* it is always important to look beyond the "truth" of science and analyze objects on display in exhibitions in relation to the circumstances of their production and the context of their presentation. Recent scholarship on museums sensitizes us to the means by which these institutions render invisible the ideological structures that underpin their meanings, and remind us of what Foucault convincingly argues: power is always entailed in the production of knowledge, and knowledge is inseparable from power.[6]

The Smithsonian's broad ambitions in relation to its ethnological collections have been examined by Curtis M. Hinsley and other scholars, but there is still a paucity of work focusing on individual objects—their manner of collection, methods of display, and specific contribution to the meaning of the larger exhibit—as well as the relationship between the bodies and artifacts displayed and the living peoples they purported to represent. Scholarly analyses of exhibitions mounted over a century ago present many difficulties: archival documentation is patchy; records about individual specimens are hazy or nonexistent; museums had a vested interest in covering their tracks and normalizing displays. Using archival documents, newspaper reports, photographic records, and published sources, this chapter analyses the commissioning, creation, and public display of the Fort Marion Indian heads. It illuminates the fundamental importance of Native Americans to the wider meaning of the Smithsonian's exhibition by showing how these plaster-cast heads of indigenous Americans—decapitated synecdoches for their "primitive" bodies—were assigned a vital role in the public construction and display of racial differences that bolstered the intertwined narratives of United States nationhood and indigenous extinction (Figure 9, photo gallery).[7]

The Fort Marion heads were enthusiastically commissioned by the Smithsonian's assistant secretary, Spencer Fullerton Baird, in 1877, as soon as he became aware that a group of Plains Indians was being held prisoner in St. Augustine, Florida:

> It has always been difficult to obtain face casts of the North American Indians. They manifest a deeply rooted aversion to the process required, and, indeed, a superstitious fear generally of being imitated in any manner, even by pencil or camera. The face masks from nature now in existence have, for the most part, been taken from the dead, with the consequent lack of vital expression, and the opportunity of obtaining life-like similitudes of 64 Indian prisoners of war, of at least six different tribes, was promptly embraced by the Smithsonian.[8]

For Baird, the chance to make face-masks from living Indians promised to advance the study of different races, where attention "has been directed very closely to the shape of the head, of the lineaments, and of the external form during life . . . and every opportunity of securing accurate casts, in plaster, of native races of a country is eagerly embraced."[9] He knew the famous Schlagintweit brothers had made 275 life-masks of Asiatic tribes in the 1830s, which had become "standard objects in the principal ethnological collections of the world," and was determined to include American life-masks in the Smithsonian's fast-growing collections.[10]

From the moment he took up his post at the Smithsonian, in 1850, Baird had used his prodigious expertise in ornithology, geology, mineralogy, botany, and anthropology to coordinate a drive to collect botanical, zoological, and ethnographic species of all kinds from across the continent and the wider world.[11] Baird's campaign was acclaimed by a contemporary biographer:

> [Baird] literally turned every branch of the government into a clearing house for the National Museum and the Smithsonian Institution; the consulates, the agents of the government, surgeons, army officers, ministers, soldiers, lighthouse keepers, revenue service, officers of the marine corps, the engineer department; no branch of the service was over-looked by this indefatigable collector who had the power to interest everyone in his work and induce them to send in animals, plants, minerals, fossils, fruits and flowers, or Indian implements from the localities in which they were stationed.[12]

Once examined, categorized, and organized systematically, Baird believed that this diverse collection of specimens held the power to reveal the workings of nature and the deeper laws governing man's place within it. When Baird secured Congressional funding and authorization to construct a building for a National Museum in Washington, the Museum's grand opening, in 1881, marked both the achievement of his thirty-year dream and the opportunity to reorganize and publically display the Smithsonian's extensive collections.

The National Museum opened at a time when a growing international group of museum professionals was starting to define the museum as a vital institution of public education, and to regularly communicate with each other about the methodology and practice of display. Their ideas would have a powerful impact on the new galleries of the Smithsonian. In England, General Augustus Pitt Rivers had developed what became known as the "typological method," founded on the principle that progression in human cultures, from simple to complex, mirrored the development of species and could be laid out pedagogically in the museum.[13]

At the Smithsonian's National Museum, thousands of industrial, mechanical, botanical, zoological, as well as anthropological specimens were laid out in standardized plate-glass cases mounted on mahogany bases. The single-storey building, constructed to a symmetrical design, allowed a series of halls, courts, and exterior-wall exhibition spaces (called ranges) to all interlock on a square footprint around a central rotunda. The building reflected in its architecture the integrated message carried by the specimens. The *Handbook* set out this point for visitors:

> The Museum is built up on a philosophic classification, intended to embrace the whole universe, and minute enough to find a legitimate place for every object. . . . There is throughout nature a steady progression in history and in structure from the simple to the complex, from general to particular.[14]

This "steady progression" observed in made objects, organized to separate the "primitive" or "savage" from the "civilized" or "advanced," was extrapolated to demonstrate a parallel gulf between the peoples that created them. As Curtis M. Hinsley argues, "the presumed wide gulf between Indian and Anglo-Saxon civilizations" was thus simultaneously confirmed and inscribed.[15]

One contemporary, Barnet Philips, an early visitor to the museum, gave a reading of the new exhibition in his letter to the *New York Times*, which indicated that he had understood the museum's key message:

> It takes man for its central pivot, and around this is to revolve everything that man has done in the past or in the present in the world he lives in . . . This museum, besides, is to enter into

every detail of human life, not only of the present but of the past, and is to be the custodian of its future . . . There is nothing, ever so trivial, which is thought unworthy of notice. The study of the evolution of anything is supposed to impart its lesson, and the spinning-wheel of a past time is to lead up by many stages to the more perfected mechanisms of today.[16]

In every "detail of human life," the enormous array of specimens presented to the American public in the National Museum was organized in sequences that always pointed forward to "the more perfected mechanisms of today." Despite the weightiness and apparent neutrality of all the physical evidence on display, it was deliberately arranged to create a clear ideological narrative that proceeded according to a sequence of steps. These progressions ran parallel to those that Barnet Philips had noted in the textile exhibit, "from the simple whorl of the stone or terra-cotta, used by savage or semi-civilized man . . . to the steam-spinning apparatus and the power and jacquard looms."[17] The repeated presentation of these patterns of progression provided the wider frame for visitors' understanding of every aspect of human society. It shaped the way all information about non-Western peoples was both communicated and received.

Objects extracted from their temporal and cultural contexts were arranged to provide evidence of a chronology of evolutionary advance. This process facilitated the disruption of linear time and the thrusting of non-Western peoples back into the prehistoric past, where they were situated as representatives of the present's pre-history.[18] When museum visitors gazed at the plaster heads of Native Americans or inspected the mannequin bodies displaying the clothing of indigenous peoples who still inhabited the continent, they were invited to believe that they were staring back down the tunnel of history; this production of so-called primitives as the living dead, has been described by Bernard McGrane as "necrology practiced on the living."[19]

Although perhaps obvious, it should be noted that everything in the museum was dead. Beyond the guarantee this gave to maintaining stability in the sequencing of objects and peoples on display, it also introduced the pervasive suggestion of disappearance or extinction, not just of "primitive" implements, but also of the peoples who made them.[20] Specimens exhibited in the museum were frozen in time forever; they and their owners had been permanently robbed of agency.[21] Unlike living displays of primitive peoples, they could not speak out to contradict the "primitivism" they were supposed to embody.[22] Classified as surviving remnants, rather than examples of living peoples and their cultures, their safe keeping was claimed by the museum. George Brown Goode, who organized the displays and was Baird's right-hand man, reported with satisfaction that "American ethnological museums are preserving with care the memorials of the vanishing race of red men."[23]

These "memorials" not only contributed to the sequence of the evolutionary narrative, but also to the creation of a system of knowledge that legitimated the national political structures on which the institution rested. The power relations that had facilitated the assembling of the Smithsonian's vast collections were invisible to the eye of the visitor, allowing the bodies and material cultures of America's internally colonized peoples to be re-contextualized as the remnants of dead or dying peoples. The plaster casts of the Fort Marion prisoners represented a single element in a much wider story of death and extinction, but their American status was of crucial importance; it allowed visitors to scrutinize at close range the "living dead" from within their own national boundaries, and thus gain an understanding of their own relationship to these "dying" peoples. The broad message conveyed by the museum, as Tony Bennett suggests, provided its American visitors "with a set of resources through which they might actively insert themselves within a particular vision of history."[24]

In the museum's early days, the Smithsonian's ethnographic displays had been unsystematic, with no particular emphasis on America.[25] In the National Museum, however, although other nations were represented, the Ethnology Department focused very strongly on the populations of America. The walls of one room exhibited over five hundred of George Catlin's paintings of tribal leaders, villages, and customs. Acquired by Baird for the Smithsonian in 1879, these were the product of Catlin's prodigious effort to make a visual record of peoples he judged to be "dying," and, as he hoped, to "thus snatch from a hasty oblivion what could be saved for the benefit of posterity."[26] His paintings contributed to the wider message of indigenous decline and disappearance, and the parallel story of America's rise.

Throughout the whole Smithsonian exhibit the presentation of displayed items, and their relationships to one another, took precedence over their particular and separate histories, with details of the provenance of individual pieces invariably absent and often lost. Many of the Kiowa, Comanche, Cheyenne, and Arapaho imprisoned at Fort Marion, whose heads were now on public display, had been powerful leaders of their communities and recognized as such when they were signatories of the Treaties of Medicine Lodge (1867). These treaties were highly controversial and subsequently these leaders had led the dissident groups who for nearly a decade held onto their old lifestyle and lands and refused to settle on reservations. Only a persistent and aggressive campaign by the army, led by General Sheridan, had brought their defeat, exile, and imprisonment. Transported to Florida in 1875, with no date set for their release, the prisoners became subject to the demands and decisions of the United States' government.[27] They lost control of their lands, their freedom, and after Clark Mills made their life-masks, even the representation of their own heads.

In 1877, Spencer Fullerton Baird commissioned this well-known Washington-based sculptor to travel to St. Augustine to make life-masks of all the Indians imprisoned at Fort Marion. Mills had achieved prominence when he won a competition and created the first, full-sized, bronze equestrian statue of Andrew Jackson.[28] He had also patented a unique method of making face-casts, and his life-sized, plaster facsimiles of a subject's head, reproduced "all the wrinkles, pimples, eyelashes, and eyebrows clearly and distinctly," as well as exact details of the contours of the face and head.[29] Mills had a special interest in Indian heads that predated his visit to Fort Marion. In 1864, he had cast "the heads of several Indian chiefs" who were visiting Washington, and, on examining them closely, was powerfully struck by what he described as "the peculiarities of the Indian head." Reporting his observations to Baird, the latter immediately "proposed to purchase the heads for the Institute."[30] More than a decade later, when Baird learnt about the Fort Marion prisoners, he would invite Mills to execute his Fort Marion project.

Clark Mills' attention to the human head was driven by his conviction that its shape and contours revealed the personality and attributes of an individual. He was strongly influenced by the work of Franz Joseph Gall (1758–1828), the inventor of phrenology, whose ideas had been disseminated in the United States by the Scotsman, George Combe. Throughout his working life, Mills sustained an active interest in studying and modeling the human head, and the terms in which he hypothesized about different cranial shapes reflected this interest in phrenology.

A popular pseudoscience before the Civil War, phrenology retained a following in the United States until the end of the nineteenth century and was inseparable from ideas about scientific racism.[31] Phrenologists believed that specific areas of the brain had very particular functions, and that the level of development of each of these functions could be ascertained by examining an individual's cranium. They thought the brain contained a range of organs that controlled separate functions and that like muscles, these organs could be expanded through exercise; any such expansion would be matched by changes in the shape of the skull. Confident that separate functions could be improved, particularly in childhood, phrenologists were equally certain that the degree of possible improvement of a race was determined by the original organization of the brain. When reflecting on the westward advance of the American population, George Combe asserted, "the existing races of Native American Indians show skulls inferior in their moral and intellectual development to those of the Anglo-Saxon race, and that morally and intellectually, these Indians are inferior to their Anglo-Saxon invaders, and have receded before them."[32]

Attention to Native American heads, as well as those of other "races," had been persistent in the United States. In the antebellum years, Samuel Morton used the measurement of Indian skulls to demonstrate that they were smaller than those of whites, arguing in his lavishly produced *Crania American* (1839) that this was indicative of inferior social and intellectual capabilities. Stephen Jay Gould, in *The Mismeasure of Man,* has revealed the faulty methodology that underpinned Morgan's principle thesis. But it was not, Gould argues, a deliberate attempt by Morton to distort the facts that led him to make some fundamental errors in the way he did his calculations. Rather, it was "an a priori conviction about racial ranking so powerful that it directed his tabulations along pre-established lines."[33] The pervasive power of "a priori convictions" that had influenced Morton's work did not diminish over time, but continued to provide a common bedrock of conjecture for all those who scrutinized human heads and used their findings to organize separate groups of mankind onto a hierarchical ladder. In 1867, research into Indian heads was given strong official support when the United States Army Medical Museum (AMM), which had been set up during the Civil War, was reorganized to include an anatomical section. Here, they immediately began a collection of Indian skulls and other body parts. The collection grew rapidly when the Surgeon General's Circular, No. 2, 1867, called on military medical officers to collect Indian crania, along with other bones and specimens of Indian material culture. Indian skulls, gathered from battlefields and exhumed from burial sites, were sent to Washington in their thousands.[34] Here the government's craniometric research program aimed to identify different varieties of mankind and rank them according to their perceived intellectual attributes.[35] Although phrenology was more concerned with identifying personality traits, it was always closely linked to developments in craniometry.

So when Clark Mills set out for St. Augustine to cast the prisoners' heads for the Smithsonian, his personal interest in head shapes can be seen as falling into a broad-based, pre-established pseudoscientific frame for analyzing and interpreting skulls. Mills' familiarity with phrenology led him to conclude that the shared "peculiarities" he had noted in the Indian heads he had cast in Washington were "due to their mode of life, which had cultivated the same organs in each head."[36] Baird too had a longstanding interest in human heads.[37] When he wrote to the Commissioner of Indian Affairs asking for permission to make casts of all the prisoners' heads, he insisted the project would enable close examination of the Indians' physical characteristics and so make an important "contribution to Indianology."[38] For Baird, understanding the Indian body was a crucial prerequisite to understanding the Indian.

At Fort Marion, Clark Mills made full-head casts of all the surviving prisoners—sixty-one Indian men, two women, and one child—with a clear expectation of what he was looking for. When inspecting the casts, he was convinced that he had "found the same organs developed as in the heads taken in Washington," and wrote to Baird to report his observations. Keen to pursue his explorations further, Mills suggested that he should go on to make more casts of other Native Americans who had benefitted from exposure to "civilization," to help establish whether America's indigenous populations were doomed to their backward craniological state, or if their heads and faculties were capable of development:

> If it would not be well to have casts taken of New York Indians, who had been civilized for a hundred years. And if we should find the same development of the head as in the wild Indians, we might then safely say for a fixed fact that they were made Indians and cannot be changed. But if, on the contrary, we find the organs called into action in the wild state have become depressed, and those which indicate civilization have become enlarged, then we might say all they require is education; they have as much brain power as the white man.[39]

There is no evidence that Mills went on to make his projected head-casts of New York Indians. However, at Fort Marion he would make the acquaintance of Captain Richard Henry Pratt, the prisoners' guard, who would later facilitate a different version of Mills' comparative phrenological experiment.

Pratt went on to found the Carlisle Indian School and act as its superintendent from 1879 until 1904. He ceaselessly voiced his opinion that Indians were fully capable of education, and at Fort Marion he set up a fortress school to instruct the younger men in reading and writing English. When the prisoners were eventually released, in 1878, he persuaded the Hampton Normal and Agricultural Institute (an institution founded in 1868 to educate the freemen) to enroll twenty-two of the younger men and accompanied them to Virginia to assist. Clark Mills had already made casts of the Hampton/Fort Marion students' heads, but when a new contingent of students from Dakota arrived at Hampton, Mills secured "a commission to take casts of the children's heads, so that when educated casts might again be taken to ascertain what change, if any, had taken place in the formation of their heads."[40] Although there is no record that Mills ever returned to Hampton to make the second set of comparative "educated casts," it is clear that his interest in Indians was driven by a conviction that their capacity for advancement could be ascertained by close study of their heads. The extraordinary physical detail of each individual face that was reproduced in his casts was crucial when Mills was planning a sculpture of a man like Abraham Lincoln.[41] When making physical representations of Indians, however, it was not their facial individuality that was most important to Mills, but rather the typological contours of their heads.

Although he pondered Indian capacity for education and advancement, Mills also, like most of his countrymen, harbored a belief that their race was doomed to disappear. After shipping eighteen packages to the Smithsonian, containing the sixty-four Fort Marion Indian casts, he reflected on the significance of his work in a letter to Baird:

> They are undoubtedly the most important collection of Indian heads in the world, and when they have become extinct, which fate is inevitable, posterity will see a fact simile [*sic*] of a race of men that once overrun [*sic*] this great country, not only their philognomes but phrenological development also.[42]

Mills thus saw himself as providing a scrupulous record of peoples destined for extinction.

At the Smithsonian, the Fort Marion Heads were made part of the Ethnology Department's display and integrated into a general account of "the races of man." Here, the same broad developmental message about progress, which framed the Smithsonian exhibit throughout the National Museum, was reiterated. George Brown Goode made sure that when objects of material culture were displayed: "The series should begin with the simplest types and close with the most perfect and elaborate objects."[43] As the department was specifically concerned to demonstrate the progress of peoples as well as the objects they produced, it also included individual geographical displays, each one confirming discrete stages of the process.[44]

Exhibit labeling and *The Handbook of the National Museum* provided an interpretive structure to help visitors understand what they saw and what they should expect to learn:

> Ethnology is the science of the races of man—man divided into nations. It can be shown in a museum only by models, pictures, clothing and other distinctive external marks of difference in races. This has been attempted here by means of photographs and prints, which will be found in wing-cases; and by casts of faces and busts illustrating racial peculiarities, especially of our redskins. Distinguishing points of races are manifested in their houses, dress and handiwork.[45]

In this short paragraph, the visitor is invited to believe that scientific authority supported the notion of separate "races of man," and further, that these "races" run parallel to national divisions. Although the museum could only display "external marks of difference in races," the clear implication being made here is that behind these lie important internal differences, which cannot be displayed. The "distinguishing points of races" could be displayed in their houses, dress and handiwork, and these were exhibited with the presumption that they matched the "racial peculiarities" of their makers' "faces." The visitor was positioned as "us" viewing the spectacle and objectification of "them," and immediately was both included and implicated in the relations of power that had been institutionalized through the museum's visual representations and displays.

Even without attention to the *Handbook,* the exhibit of the Ethnology Department carried a clear message for all visitors. The non-white "races of man" were all collapsed into a single category. Exhibited together in the same room, although the attributes of their housing and handiwork might be shown to vary, their common location signaled that they and their material cultures were united by their shared strangeness and "primitiveness." These qualities set them apart from the peoples who had created the "superior" buildings that were on display in the Department of Architecture, as well as the more "advanced" products and machinery exhibited in the Gallery of Textiles and Looms.

Archival photographs provide us with rare evidence that reveals how the Fort Marion heads were exhibited. In one image, they are displayed alongside life-sized models of dark-skinned tribesmen carrying bows, arrows, hide shields, and spears, whose strange hair-styles, unfamiliar clothes and/or naked flesh marked them as both different and alien (Figure 9, photo gallery). The combination of these features together carried a message that mingled hints of the exotic, erotic, savage, and primitive. Many of the models carried weapons and all were male, thus eliminating any suggestion of the female or domestic.[46]

A second photograph shows a slightly different display, with the heads flanked by a lone, life-sized Ainu male in a glass case that also contained bows, arrows, and models of reed houses (Figure 10, photo gallery). Indigenous to Japan, the Ainu, similarly to Native Americans, were deemed by the Japanese government to require "civilizing." Although a geographically incongruous alignment of peoples, the juxtaposition of Indians and Ainu in the exhibit served to collapse specific cultural identities in favor of a wider category or type they were deemed to share—primitive.[47]

These photographs reveal that the Fort Marion heads had not been used to top the museum's collection of mannequins displaying indigenous clothes. This is what Baird had originally proposed when he wrote to Pratt, in 1877, to make arrangements for the casts to be made. "We possess something like 100 full sets of clothing which we desire to make up into life-sized figures, but for which we lack a sufficient number of faces."[48] At that time, Baird expressed his wish that none of the mannequins should have the same face, but he appeared unconcerned that the tribes represented by the Fort Marion prisoners—Kiowa, Comanche, Cheyenne, Arapahoe, Caddo—might not match-up with the available sets of clothing. It appears that what he wanted was "authentic" Indian physiognomies in his displays, and that to achieve this he was ready to play a museum version of the classic children's game of heads, bodies and legs; much of the entertainment value of this game is created by the alignment of incongruous body parts.[49]

From the surviving photographic evidence, however, it seems clear that the heads were not placed on mannequins, nor were they integrated into the separate geographic displays representing different tribes and areas. Instead, they were lined up in rows of four inside the standard Smithsonian glass-fronted cases, which were then stacked on top of one another. Their mountings were unlike those used for conventional sculptured busts, which invariably include the neck and top part of the subject's torso. In stark contrast, the Fort Marion Indian heads were displayed to the public as they had been cast, severed at the top of the neck and attached to a square base. This style of presentation not only robbed the prisoners' plaster heads (and by implication the individuals they represented) of both power and gravitas, it also gave the chilling impression of decapitated heads, similar to war trophies.

Originally the plaster faces were white, but when displayed they had been tinted, and so they blended with all the other non-white models in the room. They were thus subtly denied any claim on whiteness, and its traditional associations with classical statues, noted by Kirk Savage.[50] Presented as a group, the majority of the heads were not displayed at eye level, as would be normal practice. Although the casts invited close inspection because of their realism and exactitude, visitors to the Smithsonian would only have been able to appreciate this detail when inspecting those heads displayed on the top rows. Proper viewing of the lines of heads displayed at ankle-level, however, would have necessitated kneeling down on the floor. The mass presentation of the heads suggests that the individuality of each of the warriors was unimportant; the heads were intended to be viewed as a generic group of Indians. In the context of the museum, this provided a powerful visual physical rendering of "them" being beneath "us," underwriting the deeper message conveyed by the whole exhibit.

Apart from their color, the Fort Marion heads lacked any of the usual markers of Indianness. So in order to communicate their racial "type" to the visitor, Sitting Bull was recruited as a signifier. A life-sized replica of his head and torso [not made from a face-cast] complete with braids, feather, and blanket, was placed on top of one of the glass cases containing the heads (Figures 9 and 10, photo gallery). None of the Fort Marion prisoners was Lakota, but for Americans, Sitting Bull was instantly recognizable. The best-known

Indian of his time and still closely associated with the killing of General George Armstrong Custer and his 7th Cavalry, Sitting Bull's bust worked as an "Indian marker" for the Fort Marion prisoners. For the nineteenth-century visitor, Sitting Bull's inclusion in this exhibit was also an unspoken reminder of the "savagery" that stood in the way of national expansion, the importance of Indian defeat to the advance of "civilization," and both the urgency and legitimacy for lands previously occupied by Indians to be absorbed into the United States.[51] Synecdoches for the bodies, cultures, and projected future of all Indian peoples, the heads of the Fort Marion prisoners communicated to visitors a strong political message that actively engaged with the message of the National Museum as well as that of the national narrative.

The Fort Marion plaster heads, which Baird had welcomed as "life-like similitudes," more closely resembled death-masks in their museum setting.[52] Disconnected from the actual bodies and physical realities of the lives of the individuals they represented—the majority of whom were still alive, living on reservations in the West—the heads had been deftly integrated into the museum's discourse of racial hierarchy and extinction. The narrative in which they were held did not point forward to the men's present lives, but rather back, to an assumed ahistoricism and primitivism. Responding to this feature common to many museum displays, James Luna noted:

> I had long looked at representation of our peoples in museums and they all dwelled in the past. They were one-sided. We were simply objects among bones, bones among objects, and then signed and sealed with a date. In that framework you really couldn't talk about joy, intelligence, humor, or anything that I know makes up our people.[53]

While Luna's observation stands true for the Fort Marion busts at the National Museum, a contemporary newspaper report, written by a white journalist for a white readership and therefore inevitably freighting many entrenched stereotypes, nevertheless carries a rare glimpse of the prisoners' response to the cast-making process; most notably the humor with which they greeted Mills' first completed cast, which was of the Kiowa leader, Lone Wolf:

> When the cast had been secured he exposed it to them, and their astonishment was great. They looked at the cast, then at the countenance of Long Wolf [*sic*] again and again, and seemed quite excited, and finally laughed as heartily as an Indian can laugh, and seemed much amused. He inquired of the interpreter what amused them so, and was informed they were laughing at the "white Indian," the cast of Long Wolf's [*sic*] features being perfectly white.[54]

While their heads were in the process of casting, the prisoners were free to make their commentary and to laugh and joke—they clearly found the sight of a powerful Kiowa leader rendered as white to be hilarious—but their voices and humor could not accompany their plaster likenesses to Washington. Having lost control over their facsimile heads, they could neither speak out against their inclusion in an exhibit which promoted white systems of knowledge, nor protest against their appropriation to uphold doctrines of Indian inferiority. More than a century later, the power and audacity of James Luna's work lies in its challenge to nineteenth-century museum displays such as these, which incorporate Native peoples as the "living dead," robbing them of both agency and humanity, and legitimizing the destruction of their lives, homelands, and cultures.

Since the passage of the Native American Graves Protection and Repatriation Act in 1990, Native peoples have begun to reclaim and repatriate some of "the objects among bones, bones among objects," that have been housed in American museums for over a century.[55] Masks and busts of Native heads, however, do not fall under the terms of the Act; they are not eligible for repatriation. Nevertheless, there is growing recognition that such representations are also part of Native American patrimony and that they too should be returned. In 2004, a Smithsonian project assembled information about some of the museum's 100 Native heads, masks, and busts, and made it available to descendant communities.[56] The Osage Tribal Museum, in Pawhaska, Oklahoma, offers one example of a tribe that, over a period of eight years (2004–2012), organized the return of a series of ten busts created from the Smithsonian's molds of their ancestors.[57] At the Smithsonian, the Clark Mills molds of the Fort Marion prisoners have survived. As recognition grows that these are not "simply objects among bones," there is hope that in the future they too will be used to create busts that can be returned to the communities where their original models' lives are still remembered and their deeds continue to be honored.

Notes

1. Interview with Kenneth R. Fletcher, *Smithsonian Magazine,* April 2008, http://www.smithsonianmag.com/arts-culture/atm-qa-james-luna.html.

2. The Kumeyaay are a coastal group of Native Americans from southwestern United States and northwestern Mexico.

3. Fletcher, *Smithsonian Magazine.*

4. George W. Stocking, Jr., *Objects and Others: Essays on Museums and Material Culture* (Madison: University of Wisconsin Press, 1985), 4–5.

5. Sharon Macdonald, *The Politics of Display* (London: Routledge, 1998), 3.

6. Michel Foucault, *Discipline and Punish: The Birth of the Prison* (London: Penguin, 1991); Ivan Karp and Steven D. Lavine, eds., *Exhibiting Cultures: The poetics*

and politics of museum display, Washington DC: Smithsonian Institution Press, 1991); Ivan Karp, Christine Mullen Kreamer & Steven Lavine, eds. (*The Politics of Public Culture,* Washington DC: Smithsonian Institution Press, 1992).

7. Fort Marion prisoners' heads, displayed with mannequins and Sitting Bull bust, c. 1890, "Costumed mannequins and display cases in exhibit," National Anthropological Archives, inventory number 09707900, Smithsonian Museum Support Center, Suitland, Maryland.

8. Spencer F. Baird, "Proceedings of the National Museum for 1878," vol. I, 201, http://www.archive.org/stream/proceedingsofuni11878unit#page/n5/mode/2up; also reprinted in Richard Henry Pratt, *Battlefield and Classroom: Four decades with the American Indian 1867–1904,* edited by Robert Ml Utley (New Haven: Yale University Press, 1964), 137.

9. Ibid.

10. For the history of these masks, see G. Armitage, "The Schlagintweit Collections," *Indian Journal of History of Science,* 24, 1, 1989, 67–83.

11. William H. Goetzman, *Exploration and Empire: The Explorer and the Scientist in the Winning of the American West* (New York: Alfred A Knopf, 1966); Donald Worster, *A River Running West: The Life of John Wesley Powell* (Oxford: Oxford University Press, 2001).

12. Charles Frederick Holder, "Spencer Fullerton Baird, Geologist," in David Starr Jordan, ed., *Leading Men of Science* (New York: H Holt & Co., 1910), 276.

13. Quoted in M. W. Thompson, *General Pitt Rivers: Evolution and Archaeology in the Nineteenth Century* (Bradford-on-Avon: Moonraker Press, 1977), 34.

14. *Handbook of the National Museum,* 1886, http://www.150.si.edu/siarch/guide/start.htm.

15. Curtis M. Hinsley, *The Smithsonian and the American Indian: Making a Moral Anthropology in Victorian America* (Washington DC: Smithsonian Institution Press), 68–69.

16. Barnet Philips, "Two Letters of the Work of the National Museum," reprinted from

the *New York Times,* February 3, 1882, in *Proceeding of the US National Museum,* 1881, Smithsonian, Washington DC, GPO, 1882, Appendix 2, http://www.archive.org.

17. "Department of Textile Fabrics," *Handbook of the National Museum,* 1886, http://www.150.si.edu/siarch/handbook/textile.htm.

18. Tony Bennett, *Pasts Beyond Memory* (London and New York: Routledge, 2004), 19.

19. Bernard McGrane, *Beyond Anthropology: Society and the Other* (New York: Columbia University Press, 1989), 111.

20. Tony Bennett, *The Birth of the Museum* (London and New York: Routledge, 1995), 46–47.

21. Anne McClintock, *Imperial Leather: Race, Gender and Sexuality in the Colonial Contest* (London and New York, Routledge, 1995), 40.

22. Andrew Zimmerman, *Anthropology and Antihumanism in Imperial Germany* (Chicago and London: University of Chicago Press), 37; Roslyn Poignant, *Professional Savages: Captive Lives and Western Spectacle* (New Haven: Yale University Press, 2004).

23. *Annual Report of Board of Regents of the Smithsonian Institution: Showing the Operations, Expenditures, and Condition of the Institution for the Year Ending 1897,* 78.

24. Bennett, *The Birth of the Museum,* 47.

25. Out of fifteen cases displaying ethnographic items, only one featured North American material that was random in both subject and organization, William J. Rhees, *Guide to the Smithsonian Institution and National Museum,* Washington DC, 1859, 70–75.

26. George Catlin, *Letters and Notes,* vol. I, 3: "Department of Ethnology," http://www.150.si.edu?siarch/handbook/ethnol.htm.

27. The decision to exile and imprison these warriors indefinitely was taken by President Grant. They were held at Fort Marion, St. Augustine, 1875–1878.

28. Cyrus Townsend Brady, *The True Andrew Jackson* (Philadelphia & London: J. B. Lippincott Co., 1906), xvii.

29. Descriptions of the process Mills used can be found in Pratt, *Battlefield and Classroom,* 136–7, and in "A Sculptor Among the Indians," July 28, 1877.

30. Clark Mills to the Editor of the *Washington Republican,* published in the *New York Times* as, "Indian Phrenology: Clark Mills on the Cranial Bumps of the Savages," May 22, 1882, http://query.nytimes.com/gst/abstract.html?res=F20816F63A 5A11738DDDAB0A94DD405B8284F0D3.

31. Andrew Bank, "Of 'Native Skulls' and 'Noble Caucasians': Phrenology in Colonial South Africa," *Journal of Southern African Studies,* vol. 22, no. 3, September 1996, 387–403; Paul Turnbull, "British Anatomists, Phrenologists and the Construction of the Aboriginal Race," *History Compass,* vol. 5, no. 1, January 2007, 26–50. The *American Phrenological Journal* ceased publication in 1911.

32. Quoted in Reginald Horsman, *Race and Manifest Destiny: The Origins of American Racial Anglo-Saxonism* (Cambridge: Harvard University Press, 1981), 58–59.

33. Stephen Jay Gould, *The Mismeasure of Man* (New York: W. W. Norton, 1981), 69.

34. Jack F. Trope and Walter R. Echo-Hawk, "The Native American Grave Protection and Repatriation Act Background and Legislative History," in Devon Mihesuah, ed. *Repatriation Reader: Who Owns American Indian Remains?* (Lincoln: University of Nebraska Press), 127. In May 1898 the AMM transferred 2,206 skulls to the Smithsonian, and many of these have since been repatriated under the Native American Graves Protection and Repatriation Act of 1990.

35. E. Juzda, "Skulls, Science, and the Spoils of War: Craniological Studies at the United States Army Medical Museum, 1868–1900, *Studies in History and Philosophy of Biological and Biomedical Sciences,* 2009, 40, no. 3, 156–67, Juzda suggests that those making government policy and researchers did not see eye-to-eye about craniological research and so interest in the collection waned.

36. Clark Mills to the Editor of the *Washington Republican,* published in the *New York Times* as, "Indian Phrenology: Clark Mills on the Cranial Bumps of the Savages," May 22, 1882.

37. See J. G Heck, *Iconographic Encylopedia of Science, Literature and Art, vol. 3* (New York: Rudolph Garrigue, 1851), translated and edited by Spencer Fullerton

Baird, 123–130, in which different classificatory systems of "types" of the human race are assessed based on skull measurement.

38. Spencer Fullerton Baird to the Commissioner of Indian Affairs, June 1, 1877, Letters Received by the Office of Indian Affairs, Central Supt., Roll 68, M234, RG75, National Archives.

39. "Indian Phrenology: Clark Mills on the Cranial Bumps of the Savages," May 22, 1882.

40. See "Indian Phrenology: Clark Mills on the Cranial Bumps of the Savages," for a description of his purpose. For full details of the 50 Hampton students whose faces were cast, see *Proceedings of the United States National Museum*, 1879, 211, http://si-pddr.si.edu/dspace/bitstream/10088/12274/1/USNMP-2_79_1879.pdf. For a report of how these students were brought from the Indian Agencies in Dakota, see *Report of the Commissioner of Indian Affairs*, 1878, 173.

41. Clark Mills, *Plaster Life Mask of Lincoln*, cast after 1865 original, National Portrait Gallery, Smithsonian Institution, NPG.71.26.

42. Quoted in Brad D. Lookingbill, *War Dance at Fort Marion: Plains Indian War Prisoners* (Norman, University of Oklahoma Press, 2006), 128.

43. Hinsley, *The Smithsonian and the American Indian*, 94.

44. Curtis M. Hinsley explains that in the department, the evolutionary displays of the Pitt Rivers Museum were combined with the Gustav Klemms geographic approach, reflecting mans' dependence on nature, 87–91, 94.

45. "Department of Ethnology," *Handbook of the National Museum*, http://www.150.si.edu/siarch/handbook/ethnol.htm.

46. Fort Marion Prisoners' Heads, Displayed with Mannequins and Sitting Bull Bust, c.1890. "Costumed mannequins and display cases in exhibit," National Anthropological Archives, inventory number 09707900, Smithsonian Museum Support Center, Suitland, Maryland.

47. Mannequin of Ainu man, displayed with Fort Marion prisoners' heads and Sitting Bull bust, c. 1890, "Costume of mannequin of Ainu man with implements and house models in display case in exhibit hall," NAA, inventory number 09708400.

48. Spencer Fullerton Baird to Richard Henry Pratt, May 21, 1877, Pratt Papers, Beinecke Rare Book and Manuscript Library, Yale University.

49. It is ironic to note that when the game was first played in France, it was known as "le cadavre exquis," or exquisite corpse.

50. For a telling analysis on the white head of "Apollo Belvedere," see Kirk Savage, *Standing Soldiers Kneeling Slaves* (Princeton, Princeton University Press, 1997), 9–10.

51. For an excellent analysis of the historical implications of the Battle of the Little Big Horn and the way in which "the losers transformed a defeat into a mythic symbol of nation-building," see Colin G Calloway, *First People* (Bedford/ St Martins, Boston, 1999), 342.

52. Spencer F. Baird, "Proceedings of the National Museum for 1878," vol. I, 201, http://www.archive.org/stream/proceedingsofuni11878unit#page/n5/mode/2up; and reprinted in Pratt, *Battlefield and Classroom*, 137.

53. Fletcher, *Smithsonian Magazine*.

54. "A Sculptor Among the Indians," July 28, 1877.

55. Fletcher, *Smithsonian Magazine*.

56. Kathryn Musica, "NMNH or Bust: Museum Sculptures of Native Americans," http://www.mnh.si.edu/NHRE/RTP2004.html#Musica.

57. "The Osage Ten: Unveiling of Margaret Goode," *Osage News,* April, 2012, http://www.osagenews.org/video/osage-ten-unveiling-margaret-goode.

4

William Lanné's Pipe

Reclaiming the "Last" Tasmanian Male[1]

LYNETTE RUSSELL

The Tasmanian Museum and Art Gallery in Hobart in Tasmania recently installed a new Aboriginal display. One of the glass cases contains a nineteenth-century tobacco pipe labeled "William Lanné's pipe." This exhibit, presented without any explanatory context, demonstrates the simplicity, the humanity, and the utter recognizability of the item, and by extension its owner. Anyone familiar with the story of William Lanné, whose body is often described in terms of being the last Tasmanian "full blood male," is gently reminded that he was much more than that tragic story; he was an individual with needs, and desires—a real breathing (smoking) person (Figure 11, photo gallery).

Hobart is also where Trucannini died in 1876. She was described as the last "full-blood" Tasmanian Aborigine and as such her death was thought to herald the extinction of an entire race of people. In the months before her death, she was reported to be gravely concerned with the possibility that her body would be dissected and displayed, and was "haunted" by the memories and images of the dissection and mutilation of her compatriot, William Lanné (1835–1869). Facing death, alone and afraid, she "begged to be protected from the body snatchers." Although assured that she would not face a similar fate to Lanné, she rested in her grave for only two years. Her exhumation, granted on the basis that she represented an important scientific commodity, was approved on the condition she not be publicly displayed. From 1904 until 1947, however, Trucannini's skeleton could be seen at the Royal Society of Tasmania Museum. Here she remained an iconic reminder of the colonial project, the "last of" her people. For decades, even after her removal from public display, Trucannini's remains could still be viewed by appointment.[2]

In 1869, seven years before the death of Trucannini, William Lanné died. At his death, Lanné was considered to be the last male member of the Tasmanian Aboriginal race, whose demise was regarded as a "melancholy footnote" to Australia's colonial past.[3] Lanné's death, and the subsequent mutilation of his body, has usurped his life's achievements, and he is today remembered primarily for the indignity of being mutilated and harvested as a scientific specimen.[4] This macabre episode in Australia's colonial history has echoed down through the ages and continues to haunt the present. The ghost of William Lanné's dismembered body figures much larger than his life. This chapter will examine aspects of the life of Lanné, giving particular consideration to his independence and the autonomy he exerted in his chosen profession of harpooner and whaler. Lanné and other Aboriginal men played important roles in the vital nineteenth-century industry of whaling. Exploration of his life achievements reveals that Lanné was not merely a victim of the colonial enterprise, but that, at least until his death, he was an active participant in the newly formed colonial society that was nineteenth-century Hobart.

On March 5, 1869, the Hobart town newspaper, the *Mercury,* announced the death of William Lanné and observed that the funeral was to take place the following day. Lanné for many years had been something of a celebrity in Hobart, and as the last so-called full blood Tasmanian Aboriginal man, his death "aroused considerable scientific attention."[5] He was only thirty-four years old. Two weeks prior to his death Lanné had come off the whaleship *Runnymede,* which had been whaling in the Southern Pacific under Captain Hill and her owner Captain Bayley. On discharge Lanné had received the balance of his wages, which was the substantial sum of £12 13s 5d. He returned to his lodgings at the Dog and Partridge Hotel, where he had lived intermittently for a number of years. His death is variously described as the result of English cholera, typhoid, choleric diarrhea and the effects of advanced alcoholism as "beer and rum had taken their toll of him."[6] The day prior to his death, he had visited a dispensary run by Dr. Smart, where he received some medication. However, he began to deteriorate and the publican sent for a Dr. Atherton, who requested that Lanné be taken to the hospital. While dressing for the short trip to the hospital, Lanné collapsed and died.

The passing of King Billy, as he was often called, was regarded as a momentous occasion, though not one of which Europeans should be proud. His death was seen as lamentable even though utterly inevitable. The interest shown in "the burial of the last of a large tribe, such as inhabited and were once possessors of this island [was] almost unprecedented."[7] Not everyone regarded the event with sadness, however, and some scientists saw an opportunity to study his remains to determine such pressing intellectual questions as the "degree of relationship between Tasmanian aboriginal people and the gorilla."[8]

The study of human remains was, at that moment, a global phenomenon, and one which "linked the old world and the new."[9] Indigenous peoples, along with criminals, paupers, and the "insane," were much more likely to be subjected to the scientific gaze and to posthumous collection and dissection. In the late nineteenth century, anthropology emerged as a discipline and the cultural varieties of mankind were sought after for study: Native Americans, Australian Aborigines, Hottentots, New Zealand Maori. Explanations for human difference were sought in dissection and comparative examination of the members of these groups. As the essays in George Stocking's *Bones, Bodies, Behavior* illustrate, there was a transnational trade in scientific ideas, methods, and indeed actual human remains at this period.[10] King Billy became part of this movement as his body was fragmented and dissected in several stages, although there is no evidence that it was ever subjected to scientific study.[11]

The fragmentation of King Billy fostered wild speculation about what may have happened to his various body parts. Rumors abound that his skull was transferred, traded, or just lost, and historian Lyndall Ryan has even suggested that a tobacco pouch was made from his skin.[12] Similarly, we will probably never know what happened to his ears, nose, arms, hands, or feet, all of which are supposed to have been removed and kept.[13]

As soon as they were notified of his death, the Royal Society of Tasmania immediately requested that they be permitted to harvest Lanné's remains. The dissection and "study" of Aboriginal people's bodies was a reasonably common phenomena during this period—part of the growing anthropological interest in human physical difference. However, the government declined to sanction any interference with the body, and suspecting that it might be tampered with, the Premier, Sir Richard Dry, sent instructions to Dr. George Stokell of the Royal Society of Tasmania that the body of "King Bill" should be protected from mutilation. The government's reluctance to allow scientists to tamper with Lanné's body possibly stemmed from a belief in his humanity, a desire to maintain his dignity as an individual, as well as an acknowledgment of his popularity with Hobart establishment. Sadly, the instructions were ignored and the following night in the hospital William Lanné's

> head was skinned and the skull carried away, and with a view to conceal this proceeding, the head of a patient who had died in the hospital . . . was similarly tampered with and the skull placed inside the scalp of the unfortunate native, the face being drawn over so as to have the appearance of completeness.[14]

This grisly and disgusting fragmentation was carried out under the guise of science. Lanné's remains were then prepared for a Christian funeral, which

in the manner of a farce proceeded the next day. The solemnity and dignity
of the event was captured in the *Examiner*:

> On Saturday afternoon the remains of "Billy Lanné," or as he was
> generally called, "King Billy," the last male Aboriginal of Tasmania,
> were committed to the grave . . . between fifty and sixty gentle-
> men presented themselves at the institution, and found all in
> readiness for the burial. Rumours had, [*sic*] got afloat to the effect
> that the body had been tampered with, and Captain McArthur, Mr
> Colvin, and some others interested in the deceased, from his con-
> nection with the whaling trade, requested that the coffin should
> be opened, in order to satisfy in their minds that the ceremony
> was not altogether a "vain show." This was done by Mr. Graves,
> and the body was seen by those who desired to see it, in the
> [mutilated] condition . . . The lid was then screwed down, and,
> at the suggestion of some of those present, the coffin was sealed
> [and] the coffin was covered with a black opossum skin rug, fit
> emblem of the now extinct race to which the deceased belonged;
> and on this singular pall were laid a couple of native spears and
> waddies, round which were twined the ample folds of a Union Jack
> specially provided by the shipmates of the deceased . . . The pall
> was borne by Captain Hill, of the *Runnymede*, himself a native of
> Tasmania, and by three coloured seamen, John Bull, a native of
> the Sandwich Islands, Henry Whalley, a half-caste native of Kan-
> garoo Island, S.A., and Alexander Davison, an American. The chief
> mourners were Captain McArthur, of the whaling barque *Aladdin*,
> and Captain Bayley, owner of the whaling barque, *Runnymede*.
> Among the mourners were nearly all the masters of vessels in
> port, and a many gentlemen connected with the whaling trade.[15]

The symbolism of the coffin decoration is powerful: the Union Jack ensign
under which Lanné had sailed, folded beneath the possum skin cloak, spears
and waddies (clubs) of his people. Hunting implements and implements of
warfare were thus linked together, with the iconic reminder of colonial power
(the flag), and all resting upon the warm cloak. In death Lanné was symbolically of both the new world and the old.

His Christian burial was attended by friends and colleagues (and no
doubt many who were merely curious onlookers), who knew him and recognized his humanity. But after death, William Lanné, harpooner and whaler,
was transformed into a scientific specimen and commodity. His body became
a resource, quarried as an example of "primordial" or first man. For some
this indignity could not be assumed to be merely on account of his "race."

The actions of scientists were abhorred and feared, as an editorial in the *Launceston Examiner* observed:

> The common law has thrown its protection around the remains of the subjects of the crown, and if Black Billy's dead body can be mutilated and purloined with impunity no departed relatives can be considered safe in the resting place to which we consign them.[16]

It is clear that these correspondents to the newspapers believed the sanctity of the grave, especially a Christian grave, should be observed regardless of the race of the individual interred. Reverend Cox, who officiated at Lanné's funeral, wrote a little over a month later in the *Evening Mail*:

> nothing [is] more loathsome than the mutilation of the unburied corpse of the poor Aboriginal, unless it be the violation of his grave immediately after the rites of Christian burial and the impunity that has been granted that profane and indecent act.[17]

Spurred on by the public outrage, the Colonial Secretary conducted an extensive enquiry into the actions of the surgeon William Crowther, his son Bingham (and others), and the role they had played in the mutilation of Lanné's body, leaving a significant archive of correspondence. In one letter Crowther argued that the Crown had no right to decide what happened to Lanné's body as he

> *was not a pauper,* when he died, with undischarged obligations to the Crown, but an *articled seaman* to the *Runnymede,* a *free man,* and one whose term of service at the time of his death had not expired, and over whose remains the executive government, though you, had not the supported control, or claim to.[18]

Crowther's statement seems rather odd given that governments generally do exercise control over the bodies of their subjects in terms of protecting them from precisely the type of fate that met Lanné. Crowther disingenuously used the term "free man," for clearly he did not see Lanné as an historical actor, but rather as an anthropological specimen, and therefore the entitlement of science. As Crowther did not, however, offer any reasons why *he* should be entitled to have access to the body, one can only assume that he was arguing for the primacy of science. The Colonial Secretary responded, stating that they: "fail[ed] to perceive why the circumstances stated . . . should deprive the Executive of the right to protect Lanné's remains or confer [on Crowther] . . . the right to deal with them in any way."[19]

The prominent role played in the funeral proceedings by whaling colleagues and the captains of whaling vessels indicates that within the industry Lanné was well-respected and liked. After viewing Lanné's body, whaleship captains Charles Bayley, John McArthur, and Charles Colvin wrote a letter to the Colonial Secretary. At the same time they also sent a letter to the *Examiner*. Both Bayley and McArthur had sailed with Lanné on their ships *Runnymede* and *Aladdin*. Using the strongest terms they noted:

> It is openly stated and generally believed that the body buried today in St David's Church as that of William Lanné has been shamefully mutilated before interment. We beg that you will issue instructions to have the body exhumed so enquiry may be made and the public mind set at rest on the subject.[20]

The whaling captains wanted Lanné's body exhumed to demonstrate that it had been interfered with and determine who was responsible, and who might be punished. After writing the letter the three captains decided to visit Premier Dry to request that Lanné's grave be "watched during the night." However, the Premier's instructions to leave Lanné interred were "in some way . . . miscarried" and the grave was disturbed and his body removed. William Lanné was unearthed less than a day after he was buried and was rendered an anthropological specimen.[21]

Although the focus has so often been on his death and the utilization of his body as an anthropological specimen, to be exploited, while alive William Lanné was also a colleague, a whaler, and a friend. The role Lanné played in the life of whaling Captain Charles Bayley, for example—and vice versa—will probably never be fully known, although there are some tantalizing clues. Captain James Bayley sailed the southern oceans with a "chequerboard crew"; his shipmates hailed from numerous places around the world, and included indigenous peoples from Australia and the Pacific. Just outside of Hobart, the National Trust today manages Runnymede, the former home of Bayley, named in honor of his favorite whaling ship. Here, amongst Bayley's personal possessions, can be found his photo albums, with two photographs (*cartes de visites*) of William Lanné. Inside the photo album are some newspaper clippings from 1869, relating to the events surrounding Lanné's death. One of these is a letter to the editor of the *Mercury*, written by Dr. George Stockell, attempting to establish that he was not responsible for the mutilation. Another is an anonymous piece signed by *"Albatross,"* and also published in the *Mercury*, on March 10, 1869. This letter asks a number of specific questions about the mutilation of Lanné's body and the roles played by various individuals, and begins with a maritime metaphor: "a great deal of mystification and cuttlefish practice is being carried on in connection with the dismemberment of poor King Billy." The use of the signature *"Albatross,"* a bird well-known to whalers, coupled with the cuttle-fish remark, suggests that this letter was written by someone associated with the whaling industry, possibly a captain. The letter-writer is also clearly knowledgeable with regard to the events. As one of only three newspaper clippings in the Runnymede photo album, it is possible Bayley himself wrote the letter; a testament to his feelings for Lanné.

These gruesome posthumous events have in many ways overwhelmed Lanné's achievements in life, so that most people know of him only through the details of his mutilation, or as the legendary "last Tasmanian male." The second half of this chapter, therefore, will not focus on William Lanné's death, but on his life—almost forgotten by history. It will explore his free will and personal agency—the decisions he made for himself and on behalf of his "people." The fact that he was predominantly regarded as one of the "last of his race" and treated as a scientific specimen should not prevent us from understanding and appreciating the complexity of his life or his humanity.

When Tasmania[22] was opened up to convict transportation in the late eighteenth century, little concern was expressed for its indigenous peoples. After decades of disease, violence, and dispossession, significant levels of de-population had occurred, so that by the third decade of the nineteenth century Aboriginal Tasmanians were assumed to be dying out, and a narrative of inevitable extinction was developed. The colonial government, arguing on humanitarian grounds, sought to round up the remaining Aborigines and send them to government-sponsored mission stations. George Augustus Robinson, builder and lay preacher, who had immigrated to Tasmania in 1824, was employed by the colonial authorities to proceed across Tasmania, in 1836 and 1837, to bring in any Aboriginal people who still lived on their traditional lands. After several attempts, some ending in unmitigated farce, Robinson was reasonably successful. However, one family managed to evade capture by Robinson, but was eventually located by an overseer of the Van Diemen's Land Company (VDLC).[23]

The VDLC was essentially a form of private colonization. The company intended to produce fine wool in the Australian colonies for export, and to achieve this end they established a settlement in the north of Tasmania. Burghley, as this settlement was called, experienced a number of significant attacks and what the VDLC termed robberies, but by December 1842, the VDLC overseer had secured the remaining "natives."[24] The family had lived at Cape Grim, near the Arthur River, and consisted of a husband, wife, and five children. They were sent to the Flinders Island Aboriginal settlement of Wybalenna. One of the boys was seven-year-old William Lanné. William, who later became known as King Billy, was the only one of his family to survive Flinders Island. At the Wybalenna settlement young William would have been given a rudimentary education and taught to read and write via bible studies. In 1847 Lanné attended the Hobart Orphan School, where he extended his reading and writing skills. The Hobart orphan school was intended as a place where Aboriginal children could be taught to adjust to non-indigenous society and be trained for "helpful" jobs as servants and laborers for European Tasmanians. Lanné left the school in 1851 and went to sea, apprenticed to a whaler.[25] In order to cut the costs of running the Aboriginal settlement, which had shifted to Oyster Cove on the main island of Tasmania, in 1855 the colonial government ordered that able-bodied men and all mixed-descent people were to work away from the reserve. The latter were also expected to integrate into the non-indigenous community. After this date we can be certain that Lanné was whaling on the ships *Aladdin*, the *Jane*, and the *Runnymede*.

Life onboard ship for Lanné and his shipmates was most likely organized around duties and skills, taking no account of the issue of "race." Dining on these colonial ships was not segregated, as on some American ships, and as the voyages were relatively short sojourns from Hobart out into the Pacific and southern oceans, the ships would have been well provisioned. The *Runnymede* log shows that she whaled in the waters of the Tasman Sea, Bass Strait, the southern ocean south of west Australia and south Australia, and the South Pacific east of New Zealand.

Ship life would have also been tempered by the presence of Captain McArthur's wife and family. Mrs. McArthur delivered one of her children onboard the whale ship *Aladdin* while off the New Zealand coast. Like many women who traveled on whale ships, Mrs. McArthur would have been in charge of a number of activities. Amongst her duties she may have insisted that the men observe the Sabbath, as this was generally the duty of the Captain's wife. As Joan Druett's definitive study of women aboard whaling vessels between 1820 and 1920 shows, the crew's spiritual well-being was often the concern of these whaling wives or "petticoat whalers." Under the women's influence dinnertime was usually preceded by saying grace. Evenings were spent repairing sails and other things, as well as relaxing, singing, and the occasional dance.[26] For Lanné this must have seemed a world away from the disease and despair, rations and regulations of the Oyster Cove settlement.

While onboard the *Runnymede* in 1858 Lanné would have witnessed the events that resulted in three seamen drowning. During a fishing trip to replenish the food stocks off the Solander Islands in southern New Zealand, three sailors were lost. The men had set out in a whale boat which had hit rough seas and overturned. When the ship's second mate was sent to search for them, he found the overturned boat with two men on top, clinging to the hull, while the other three had drowned. Like all sailors, whalers and others who worked the maritime industries, Lanné would have known that this was a "most dangerous occupation."[27]

Unlike the other Aboriginal people who remained at the settlement, William Lanné had significant freedom within the colonial society of Hobart.

He was free to come and go from Oyster Cove, something he was probably encouraged to do, especially as he was financially independent. He frequently resided at the Dog and Partridge Hotel, where many whalers stayed when they were in port. Like many seamen of the time, alcohol, especially rum, played a significant role in his deteriorating health. Ships frequently gave out drams of rum with meals, and fresh water was often much more difficult to get than alcoholic beverages. It has even been speculated that many of the whaling ships would have had crews that were more or less permanently alcohol-affected.

As the last of an assumed to be soon-extinct people, Lanné was locally renowned and something of a celebrity. He was favored by Governor and Mrs. Browne as an example of a civilized "Native," and his apparently good humored personality ensured he was popular amongst those who met him. James Bonwick observed that he was

> the youngest and handsomest of their tribes, they [the old ladies of the settlement] were loud in their praises of him to me. Consorting with the Europeans, and having no mate among his colour, it is not remarkable that he should have found fault with the excellent photograph . . . [by colonial photographer Woolley, whom he told] that it was "too black."[28]

It is simplistic to assume that the comment about the image being too black is an example of Lanné rejecting his Aboriginal heritage and a plea for acceptance in the white world. Such sentiments reflect also the desires of the colonists that Lanné might find a place within the colonial society, to accommodate and assimilate to European ways.

Perhaps it was Lanné's disposition that made him "a favorite," or perhaps it was simply that as the purported "last" surviving male of his people he was not perceived as a threat to the colonists. Regardless, he was accepted and even admired by many of Hobart's establishment. He was ascribed royal status, something he seems to have embraced and promoted. In 1866 the *Examiner* reported that William Lanné,

> who has been located at the Oyster Cove establishment . . . is about to visit England and his ambition is to have the honour of an audience of Her majesty Queen Victoria. He writes in a fair hand, can read well, and may be said to be intelligent. He speaks highly of Governor Browne, and Mrs Gore Browne and prides himself and his four country women at the Birthday Levee and Ball. He is also not a little proud of the several photographs in which he figures, and seems to think himself "every inch a king." His reminiscences

of "life at Oyster Cove" are of a somewhat jovial character, and Mr Dandridge the superintendent of this establishment, has won golden opinions on the part of "his majesty." Plenty of rations, boating amusement, and sundry out door healthful occupations are themes on which he admiringly dwells. No doubt "King William" would be well received in England.[29]

Although it appears that Lanné never traveled to England, he did get the opportunity to meet with British royalty when he was, along with the Duke of Edinburgh, a guest of honor at the Hobart Regatta. Historian Lyndall Ryan describes this as his "proudest moment."[30] However, such assertions more plausibly reflect the understandings of his nineteenth-century European contemporaries rather than Lanné's own perception. In 1868, during the Duke of Edinburgh's visit to Hobart Town, Lanné was introduced as the "king of the Tasmanians." James Bonwick, who witnessed the events, wrote that Lanné,

> clad in a blue suit, with a gold-lace band round his cap . . . walked proudly with the Prince on the Hobart Town regatta ground, conscious that they alone were in possession of royal blood.[31]

The regatta was an important event in the Hobart social calendar, and the fact that Lanné played a prominent role in it suggests he really did have favored status amongst the European elites. Leslie Norman describes the events thus:

> When the great day arrived His Royal Highness was marshalled from Government House to the regatta ground by a procession of dignitaries, all mounted on chargers; plus police, fire engines, the band of the Volunteer Artillery, together with banners and emblems galore. The Prince, seated in the State carriage with outriders, and accompanied by the Governor, must have had hard work to repress his smiles as the procession wended its way through the city. On arrival at the regatta grandstand, he was met by "King Billy . . . and Queen Trucaninni."[32]

William Lanné's life is difficult to fully apprehend; he was a complex figure living in difficult times. Wandering around the banks of the Derwent River in his blue suit and gold-trimmed cap alongside British royalty, it would be easy to see Lanné as a dupe: a colonial parody. While some of the colonists may have seen him as such, this is again too simplistic. William Lanné, by virtue of his personality and social skills, exercised an unprecedented degree of personal agency and subjectivity. His economic independence, though initially forced on him by the managers of the Aboriginal station, enabled him to come and go as he pleased. Furthermore, whether from his whaling experience or simply as a natural aspect of his personality, Lanné was also a leader, in the sense that he took on responsibility for the care and concern of the women who remained at the Oyster Cove settlement. He often operated as their go-between. It is unlikely that the colonists considered him to be a "king" in the manner of English kings, but he was undoubtedly a leader, even if the people he was leading were significantly disempowered, reduced in number and, in the mind of the colonists, fast approaching extinction.

Further evidence to support the idea that William Lanné was not merely a mockery of assimilation comes from his commitment to what he referred to as "my people." As a "king" clad in his royal blue suit or, alternately, his whaling gear, Lanné could also act mimetically to disrupt colonial society, and in particular the legitimacy of the colonial government. He might well appear to be thoroughly Europeanized, but he was not. As he had been absent from Oyster Cove through most of the 1850s he had not suffered the disease, desolation, and hopelessness which afflicted many of his "people," but he made many complaints about the living conditions of those that remained on the Government Station.[33] His complaints were relatively well-received, and the colonial governments attempted both to rectify the problems at Oyster Cove and also to placate Lanné. Their reaction to him suggests that he was far more that a colonial puppet, and as an active participant in colonial Tasmania he was someone of whom the authorities took notice.

Matthew Kneale, in his award-winning novel *English Passengers,* renders in fiction something which many historians have also implied: namely, that the Anglo-settlement of Tasmania gave no opportunities to Aboriginal people, and all that was expected of the indigenous population was to "die quiet and smiling and thankful."[34] William Lanné, in his short but eventful life, proves such assertions wrong. Lanné's body was consumed by science in death, but in life he exerted his autonomy and sought out the opportunities that colonialism offered. By continuously focusing on the posthumous events of dissection and mutilation, previous generations of scholars have contributed to the perception that Lanné was merely a victim of the colonial enterprise. While such an assertion clearly has merit, it is by no means a complete picture. The process of colonialism and the subjugation of indigenous people is neither complete nor unidirectional. While much has been written about the military resistance Tasmanian Aborigines mounted, and the so called "black wars," there has been little focus on other forms of reaction to European colonization, in particular the ways Aboriginal Tasmanians engaged with and make use of social and economic opportunities that arose.

William Lanné exploited these opportunities in various ways. As a whaler he found economic and personal freedom. As the "darling" of colonial society, he could be seen as a colonial parody or a powerful mimetic, but equally, he might well be regarded as being well-adjusted and partially assimilated to white society. We should not diminish the horror of his mutilation nor should we disregard the impact European invasion and colonization had on Aboriginal Tasmanians more generally. Lanné's life has all the hallmarks of tragedy: an early demise; a childhood fractured by colonial invasion; loss of family members; no known descendants. Generations of scholars have lamented his passing as the "last" of his kind and reduced his life to its posthumous horror, characterizing him as the ultimate colonial tragedy; a man who thought himself equal with the colonizers yet who clearly was not. Yet, as this chapter has shown, there were also glimmers of freedom, autonomy, accommodation, and even friendships in his life.

Aboriginal people affected colonial society by various means and William Lanné—the man, the whaler—affected those who knew him in myriad ways. In an effort to reverse the process whereby Lanné the historical figure became Lanné the anthropological specimen, this study has taken the mutilation of Lanné's body as the starting point to retrospectively look back at his life. From the bloodied table of Crowther's morgue, William Lanné emerges as someone who in life was respected and liked, and who in death split the colonial society open, revealing that the racist discourse of science could not be used to explain away what was regarded by many at the time as inhumane actions. Colonial society attempted to exert control over him in death, via his body. But in life, his freedom at sea is ultimately unrecorded and beyond our gaze. There he lived a free man, and will continue to do so in our imaginations, with his shipmates beyond our reach and in dignity.

Postscript: The Dog and Partridge Hotel in Hobart still stands. Today its front rooms house an opportunity or thrift shop, the back of the hotel building is the offices of the Hobart City Mission. In 2007, I was shown through the building by the Mission's Director, Dennis Cousens. Up the rickety ancient stairs we entered the once-residential rooms, boards now covered the tiny fireplaces, but the communal bathrooms and a kitchen were intact. The rooms had been modified little to accommodate modern offices—a wall removed. Each room's dimensions could still be defined. Although we did not know which room was Lanné's, this was the place where he took his last breath; where, when on land, he lived, ate, drank, slept. Like his pipe in the museum's glass case, this represents William Lanné the man, the whaler, the friend and relation. It reminds us to never lose sight of the fact that it he was a man, who was turned into an "indigenous body."

Notes

1. For a more extensive exploration of the issues, see Lynette Russell, *Roving Mariners: Australian Aboriginal Whalers and Sealers in the Southern Oceans, 1790–1870* (SUNY Press, 2012).

2. Ian Anderson, *Reclaiming Tru-ger-nan-ner: Decolonising the Symbol in Speaking Positions: Aboriginality, Gender and Ethnicity in Australian Cultural Studies* (Melbourne: Victoria University Press, 1995), 31. Needless to say the Tasmanian Aborigines are certainly not extinct, as evidenced by the vibrant and politically active Aboriginal community of Tasmania. See Lloyd Robson, *A History of Tasmania, Volume 11: 1856 to 1980s* (Oxford: Oxford University Press, 1990), 6–7.

3. W. E. H. Stanner, *After the Dreaming: The 1968 Boyer Lectures* (Sydney: ABC Books, 1968), 56.

4. See, for example, Tim Murray, "The Childhood of William Lanne: Contact Archaeology and Aboriginality in Tasmania," *Antiquity* 67, no. 259 (1993), 504–519.

5. Lyndall Ryan, *The Aboriginal Tasmanians* (St. Leonard's: Allen & Unwin, 1996), 214.

6. *Examiner*, 11 March 1869, 1.

7. *Tasmanian Times*, 27 March 1869, 3, col. 4.

8. Murray, "The Childhood of William Lanne," 514.

9. Helen MacDonald, *Human Remains: Episodes in Human Dissection* (Melbourne: Melbourne University Press, 2005), xii.

10. George W. Stocking, Jr., ed., *Bones, Bodies, Behavior: Essays in Behavioral Anthropology*, History of Anthropology series, vol. 5 (Madison: University Wisconsin Press, 1988).

11. Stefan Petrow, "The Last Man: The Mutilation of William Lanne in 1869 and its Aftermath," *Aboriginal History* 21 (1997), 108.

12. Ryan, *The Aboriginal Tasmanians* (St. Leonard's: Allen & Unwin, 1996), 217.

13. *Examiner*, 11 March 1869, 1.

14. *Mercury*, 27 March 1869, 3, col. 3.

15. *Examiner*, 27 March 1869, 2, col. 4–6.

16. *Examiner*, 18 March 1869, 2, col. 3–4.

17. *Evening Mail*, 10 April 1869, 2, col. 3.

18. Surgeon William Crowther's letter of 5 April 1869, AOT CSD 7/23/127, Crowther Collection, State Library of Tasmania, Hobart. Original emphasis.

19. Letter from the office of the Colonial Secretary and signed by Sir Richard Dry, 5 April 1869, AOT CSD 7/23/127, State Library of Tasmania, Hobart.

20. *Examiner*, 11 March 1869, 2, col. 5, in Crowther Collection, CSD 7/23, State Library of Tasmania, Hobart.

21. For a more detailed narrative see Helen MacDonald, *Human Remains: Episodes in Human Dissection* (Melbourne: Melbourne University Press, 2005), 136–82.

22. Tasmania was known as Van Diemen's Land until 1856.

23. For details see N. J. B. Plomley, ed., *Friendly Mission: The Tasmanian Journals and Papers of George Augustus Robinson, 1829–1834* (Hobart: Tasmanian Historical Research Association, 1966).

24. Murray, "The Childhood of William Lanne," 507–508.

25. Ryan, *Aboriginal Tasmanians*, 210.

26. Joan Druett, *Petticoat Whalers: Whaling Wives at Sea 1820–1920* (Hanover: University of New England Press, 2001), 124.

27. Log of the *Runnymede*, CSD 4/77/231, Archives of Tasmania, Hobart. See also Susan Chamberlain, "The Hobart Whaling Industry 1830 to 1900" (PhD thesis, La

Trobe University, 1988), 131; and Peter Mercer, *A Most Dangerous Occupation. Whaling, Whalers and the Bayleys: Runnymede's Maritime Heritage* (Hobart: 40° South, 2002).

28. James Bonwick, *The Last of the Tasmanians: Or, The Black War of Van Diemen's Land* (London: Sampson, Low, Son, and Marston, 1870), 394.

29. *Examiner*, 15 November 1866, 2, col 5.

30. Ryan, *The Aboriginal Tasmanians*, 214.

31. Bonwick, *The Last of the Tasmanians*, 395.

32. Leslie Norman, *Pioneer Shipping of Tasmania: Whaling, Sealing, Piracy, Shipwrecks etc. in Early Tasmania* (Hobart: J. Walsh, 1938), xx.

33. Ryan, *The Aboriginal Tasmanians*, 214.

34. Mathew Kneale, *English Passengers* (London: Penguin, 2000), 279.

III

GENDER AND SEXUALITY

5

Sodomy, Ambiguity, and Feminization

Homosexual Meanings and the
Male Native American Body

MAX CAROCCI

> The travelers anxiously regarded the upright, flexible figure of the young
> Mohican, graceful and unrestrained in the attitudes and movements of
> nature . . . Heyward . . . openly expressed his admiration of such an
> unblemished specimen of the noblest proportions of man.
>
> —James Fenimore Cooper, *The Last of the Mohicans*, 1826

Whether explicit or hidden under the appearances of gentlemanly admiration, as expressed in the quote from the novel by James Fenimore Cooper cited above, homosexual and homoerotic meanings have informed, directly or indirectly, much of the ideological and moral attitudes toward American Indians since the beginning of colonization.[1] In the following essay I intend to highlight the significance of homosexual themes, often dismissed as marginal to regimes of colonial power that, time and again, underpinned the ambivalent relationship between Native Americans and Euro-Americans throughout history.

Euro-Americans' fascination with American Indian bodies, their real or imagined (homo)sexualities, and what they came to represent are here examined to maintain that colonial and imperialist ideologies were greatly sustained by the circulation of these themes. This exploratory evaluation of imaginings, projections, and constructions shows how Native Americans are limited to being objects of fantasizing with no substantial rooting in the realities of living people. I will therefore concentrate on the meanings

69

attributed to representations of Native American male bodies to argue that Euro-American men, and particularly gays', perceptions of Native Americans echo, some would argue, colonialist or plainly racist notions associated with non-European bodies.

The examination of some of the themes at the core of this largely unwritten history combines my involvement in the gay circuit and my work with American Indians since 1989. Conversation with Native American gays, my observations of their interactions with Euro-Americans in a number of U.S. cities, and an analysis of the iconographic and popular material currently consumed in gay networks constitute the bulk of the data gathered here. Two major typologies of representations can be recognized in the overall picture of these representations: the degenerate and the noble savage. These two tropes irrevocably shaped the discourse of homosexuality upon which rest perceptions and fantasies about the male Native American body.

Amerindians and Deviance

Scholars of Foucaultian persuasion maintain that the notion of the "homosexual" is a recent invention.[2] Cognate notions that predate this term however had a significant impact on constructing the Amerindian body as deviant. In the following discussion gender deviance, sodomy, unnatural practices, and perversions are examined alongside notions of impotence, lack of virility, and femininity as a way of discussing how these related terms frame an intricate web of recurring allusions, metaphors, and iconic markers associated with the Amerindian body. These connections whilst contributing to hegemonic projects of imperial expansion, also silently coded the body of the Indian as quintessentially Other, a process that over time offered a set of meaningful referents for subordinate classes of people such as early homosexuals, gay men, and more recently, transgender people.[3] While the variety of categories of deviance common among Europeans in their descriptions of Indians often are only hinted at, it is the complexity of assumptions, insinuations, and projections of sexual and gendered meanings on the Amerindian body that framed a discourse about homosexuality which contributed to produce what Foucault called the "utterly confused category."[4]

The perplexity and bewilderment with which Europeans tried to make sense of the newly encountered peoples is reflected in contrasting perceptions of the Amerindians that over time appear at once savage yet noble, or handsome and yet brutish.[5] These contradictions, never entirely resolved, tell us about Europeans' confusion whilst at the same time offer some clues about the incipient discourse about Amerindians' sexual otherness that can be teased out by the few early pictorial references we have.

One of the first European images (Figure 12, photo gallery) depicting the abominable vices of indigenous Americans produced by Europeans is a woodcut depicting (South) American cannibals by Johann Froschauer probably printed in Germany between 1505–1506.[6] Within the frame, two men stare into each other's eyes whilst one of them touches his companion's shoulder. Given early colonial association between cannibalism and sodomy, it is reasonable to suggest that this scene might be the first visual hint to homosexuality among Amerindians.[7]

European notions of "sodomy" developed in the Renaissance played a significant part in the way in which Amerindian masculinities were perceived in this early colonial context.[8] Historian Jonathan Goldberg remarks that Renaissance Europeans were obsessed with guarding bodily boundaries. What we could call the "impenetrable" male Christian body stood in stark contrast to the naked, pierced, and therefore open body of the Indian that inverted the natural order. For this reason, the Amerindian body, although often described in colonial literature as beautiful and well proportioned, was doomed to be punished for allowing itself to be penetrated.[9] Projections of these anxieties on the Indian body constructed it as quintessentially passive, and ready to be possessed by Europeans who could then legitimately dehumanize it, enslave it, and ultimately rob it of its masculinity.

A remarkable scene of the brutal punishment of Panamanian Indian sodomites fed to dogs by Spanish Vasco Nuñez de Balboa in the early sixteenth century produced by the Flemish Theodore de Bry (1525–1598),[10] summarizes European apprehensions about Amerindians' moral conditions. His visual references to hermaphrodites' ambiguous gender contributed to speculations about their nature, and sustained arguments about the sodomitic nature of the Natives.[11] The suggestion that depictions of ambiguous, deviant Amerindians served political aims may be applicable to later textual and visual representations of feminized Amerindian males.[12]

European and North American artists used a variety of visual languages that conveyed implicit references to Amerindians' sexual ambiguity. For example, the diffusion of allegorical images of America wearing feather skirts. This invented garment came to be used in popular images and fine arts as the marker for American peoples irrespective of sex (Figure 13, photo gallery[13]).

This convention blurred gender boundaries contributing to the general perception of male Amerindians as either feminized, or sexually ambiguous. That Amerindian males were less than men or effeminate confirmed that they belonged to a different order of things.[14]

Baroque representations of androgynous Indians dancing in Arcadian landscapes visually augmented Voltaire's (1694–1778) idea that American peoples' lack of body hair was the reason why they never attempted a rebellion.[15]

Whilst Amerindians' lack of aggression was translated in artistic depictions of gentle characters from pastoral epics, Abbé Raynal's (1711–1796) conviction about the interdependence between lack of beard and virility relegated indigenous Americans to being impotent degenerates with no interest in women.[16] By contrast, Raynal's contemporaries such as the writer Michel Guillaume John de Crèvecœur (1735–1813) revealed less than feminine Natives:

> Their blankets of beaver skin fell off their shoulder, revealing their mighty chests and muscular arms . . . a painter could have drawn bodies that were perfect in proportion, limbs controlled by the muscles lightly covered with a kind of swelling that was unknown to the whites, and which among the Indians attests to their vigor, strength and health . . . this meeting of almost naked men . . . presented to the eye an impressive spectacle, and to the mind, a fruitful subject for reflection.[17]

This excerpt from Crèvecœur's published travel diaries shows that homosexual meanings engendered a dynamic tension between homoerotic subtexts and the direct feminization of the Indian proposed by Enlightenment philosophers. Such tension contributed to frame this imagined figure along the two extremes of masculinity and femininity using homosexual meanings to convey an image of Otherness. Whether the Indian was feminized or hyper-masculine, these parallel discourses similarly constructed the Indian as exotic and alien. While evoking familiar themes, from classical statues to monstrous races, this discourse was as much the product of psychological projections as it was a literary construct with important political resonance.[18]

Visual and textual references to Amerindians' hairless body sustained arguments about their inadequate masculinity or their feminine nature. Removal of body hair common among Amerindians[19] constituted sufficient evidence for Europeans to develop scientific theories that justified what was described as "degenerate" virility. The enlightened Dutch philosopher Cornelius de Paw (1739–1799), alongside the coeval French naturalist Georges-Louis Leclerc comte de Buffon (1707–1788), and the Scottish proto-anthropologist Henry Home (aka Lord Kames) (1696–1782) concurred on the idea that the "degeneration" of American peoples was evident in their body, their purportedly small genitals, and sexual behavior; namely a lack of heterosexual drive, the existence of sodomy, and other unnatural practices such as transvestism discussed by contemporaries such as the missionary Joseph-François Lafitau (1681–1746).[20]

Native American males' sexual and gender ambiguity seeped into later representations in both subtle and deliberate ways. The painting *Joseph Brant—Thayendanegea*, the leader from the Mohawk painted by George Romney (1734–1802) in 1776, shows how seemingly contradictory extremes of masculinity and femininity that framed Enlightenment's discourse about the Indian converged in the visual languages of late European Baroque. Here the leader's authoritatively masculine pose of classical portraiture is diffused with an aura of femininity given by a harness that produces creases and shadows resembling a woman's breast.[21]

Particularly significant in the Romantic period are the warrior-turned-maternal painted by Eugène Delacroix (1798–1863) in *Natchez* of 1835, and Atahualpa the Inca suzerain in John Everett Millais' (1829–1896) *Pizarro Seizing the Inca of Peru* painted in 1846.[22] In Delacroix's painting a kneeling Indian man reveals his feminine side in the tender embrace of a newly born infant that contrasts contemporary versions of European masculinity. In Millais' work the feminization of the Indian is achieved by the skirt-like cloth covering Atahualpa's hips, his recalcitrant pose resembling the avoidance of a sexual advance, and his face resembling the woman at the right bottom corner of the canvas.

Feminization, homosexuality, and degeneracy became intimately interwoven also in medical and psychoanalytical discourse to the extent that the word degenerate became synonymous with sodomite and homosexual. By mid-nineteenth century, the idea that same-sex desire depended on body characteristics was replaced by a particular attention to the mind that led to the emergence of the notion of "homosexuality" and its twentieth-century medicalization.[23] Native Americans demonstrated the psychopathologies theorized by Euro-Americans with regard to sex and gender dysfunctions.[24] It is not before the beginnings of the homosexual movement with the Stonewall riots in 1969, however, that positive evaluations of the association between American Indians and homosexuality started to appear.

Alluring Native American Bodies

Although Euro-American preoccupation with Amerindian male-to-male sexual practices served colonial objectives, it slowly lost its ideological impetus to changes in the political climate. By the nineteenth century what remained of the discourse of degeneracy was a concern with Amerindians' nudity and the application of the term degenerate to males attracted to other males.

Amerindian nudity had captured Anglos and Iberians' attention since the early phases of colonization.[25] Later commentators, by contrast, were more concerned with the aesthetics of the body than with moral messages. Although the feminization of the American Indian in later representations occurs in subtle ways, it retains a strong ideological power in delivering a message of subordinate, passive, and dominated subject at the mercy of the European gaze.

Between Baroque and Romanticism Euro-Americans often appreciated Amerindian males' bodies. Painter Benjamin West (1738–1820), in the *Death of General Wolfe* (1770–71), for example compared his Mohawk warrior with the classical statue *Apollo Belvedere* (in the Vatican Museums in Rome), then the epitome of male beauty.[26] Few years after West, George Catlin (1796–1872) excitedly described in classical terms a Choctaw ball game he witnessed between 1832 and 1839:

> . . . hundreds of Nature's most beautiful models, denuded, and painted of various colours [are . . .] a school for the painter or sculptor, equal to any of those which ever inspired the hand of the artist in the Olympian games or the Roman forum.[27]

Homoerotic subtexts run seamlessly in both European and American arts and literature throughout the Romantic period. On both sides of the Atlantic gentlemens' repressed desires, Tim Fulford reminds us, could only be lived vicariously through the intercession of fantastic scenarios and extreme sensual experiences projected onto "un-gentlemanly foreigners."[28] In the book *Love and Death in the American Novel* Leslie Fiedler takes the work of James Fenimore Cooper (1789–1851) as symptomatic of this process. In his view, the writer's depiction of the sexless relationships between American Indians and Euro-Americans reveals the homoerotic theme central to the myth of American culture.[29] In the remoteness of woods, where women are banned from boyish dreams, the Indian replaces female figures sublimating a homosexual fantasy that overdoes any heterosexual passion.[30] Not surprisingly, in Cooper' novels the white hero is inevitably coded masculine, whereas the succumbing, doomed Indian can be read as the subordinate feminine exotic that lives at the fringes of civilization.

Literature and popular discourse's archetypal Amerindian is tinted with exoticism, because it is constructed around all that is mysterious, abominable, inscrutable, transgressive, and forbidden. These features are essential characteristics that imbue with psychosexual power the object of fantasizing in a process that ultimately maps femininity upon the exotic.[31] Nineteenth-century attention to the Amerindian male nude contains much of this exotic subtext. Whether to signify closeness to nature or to evoke imagined primeval pasts lost to the encroachment of civilization, photographs depicting Native male nudity condensed Victorians' apprehensions about their own bodies whilst potentially offering titillatingly exotic pictures with homoerotic connotations to audiences of a burgeoning homosexual subculture. The ambivalence inherent in the Indian male's nude body resided in the multiple possibilities of interpretation and fruition. It simultaneously confronted heterosexual males with their own anxieties and faced heterosexual females with the possibility of desiring a non-European body that threatened both patriarchal and ethnocentric orders.[32]

Victorians' exposure to images of non-European nudes reveals power asymmetries that objectify, fetishize, and stereotype difference.[33] As Native American visual historian Aleta Ringlero remarks, Victorian portraits of naked indigenous women offered a disguised form of erotic frisson that reflects a process of objectification of the Other through the visual domination of the Indian by means of photography.[34] The male body was equally as objectified and subject to visual scrutiny as the female one because the conjunction of indigenous and feminine has for long been at the core of colonial and imperialist projects.[35]

After the Indian wars (c.1890) the feminization of the Indian is predicated upon the dyad winner-loser. Natives' military defeat put them in a position of inferiority and passivity, and representations thereafter carried established Eurocentric messages of superiority that legitimized masculine rule over females, children and subordinate classes of peoples including non-Europeans. The cowboy and Indian theme that appears in twentieth-century visual culture mirrors the winner-loser dyad which is implicitly articulated upon the homosexual contrast of active and passive sex roles. Using the established trope of the losing Indian and the victorious cowboy visual repertoires subtly convey Indian feminization. This powerful trope has tinted every aspect of popular culture from veiled allusions to homoerotic friendships between the TV characters Tonto and Lone Ranger, to imagery that explicitly rendered co-terminus the active-passive opposition with the winner-loser dyad. Gay cartoon artist Harry Bush exemplifies this trope in a drawing of a Euro-American "playing Indian" who perplexedly accepts the jubilant riding cowboy on the background of a patriotic American flag. This juxtaposition unmistakably fixes meanings associated with passivity and activity onto political scenarios that visually adds to established tropes alarming racist overtones.[36]

Early cinematography's reconciliatory tone often slipped in homosocial or homoerotic subtexts,[37] yet popular culture produced after the second half of the twentieth century glorified a type of heterosexual masculinity through the vigorous promotion of representations that subordinated Indians to their European colonizers through the playful language of leisure and domestic design.[38] Erotic and explicitly homosexual meaning however also continued to lurk behind the playfulness of subordination, during the Cold War.

Internationally renowned gay American artist George Quaintance (1902–1957), active in the post-World War II period re-balanced the relationship between Indian passivity and Euro-American activity by producing a number of artworks in which Indians are often equally as active as colonizers. Whilst his work constructs the Native American body as desirable as any other, it does it within the canons of classic European aesthetics. Native

Americans' idealized figures have little or no resemblance to the proportions of real Native American bodies.[39] These ingenuous attempts to grant visibility and status to the Indian body in the gay imagination enabled an expansion of Euro-American geographies of desire. Gay mail-order magazines such as *Athletic Model Guild, Young Physique*, and *Physique Pictorial* subsequently constructed "Indianness" by way of stereotyped props such as feather bonnets, spears and loincloths, to exoticize semi-naked Euro-American men against rather dull backdrops.[40]

Amerindian bodies continued to fuel Euro-American gay fantasies throughout the sixties, significantly, in Richard Amory's (1927–1981) *Song of the Loon* published in 1966 to which a film followed. In this book scenes of pastoral love between Indians and lonesome mountain men further contributed to fossilize Native Americans in representations of passive noble savages close to nature that resonated with James Fenimore Cooper's exotic Indian characters.

Euro-American gays' social revolution partially inspired by Native American examples of sexual acceptance did not manage to dislodge historically entrenched stereotypes. Civil rights activist Harry Hays's (1912–2002) studies of Native American *berdaches* (now Two-Spirits) brought to the public attention representations of "homosexual" Indians that, whilst closer to lived experience, were soon co-opted by anthropologists whose research on gender roles hardly produced any new insights about Native American bodies.[41]

The decline in the production of mail-order magazines in favor of widely available and more explicitly sexual publications at the end of the sixties resulted in the almost total absence of reference to American Indians. They however continued to appear in erotic scenarios made for a thriving cartoon industry that developed following to the success of the artist Tom of Finland's (1920–1991) erotic drawings. To this very day, American Indians appear in cartoons and in stories written for an adult gay public that avidly consumes popular products with an "exotic" flavor.[42]

After the seventies American Indians bodies and homosexual meanings became associated by way of icons such as the singer Cher, and Felipe Rose of the group Village People who capitalized on their indigenous heritage to break into the limelight using their bodies as stages upon which Native American meanings were performed. Once again the Native American body became coded as feminine by association with a woman and an openly gay man. Today Native Americans are associated with homosexual meanings in gay representations that show the resilience and crudeness of stereotyping.[43]

More recently, books such as the critically acclaimed *The Man who Fell in Love with the Moon* by Tom Spanbauer (1992), or films such as *Big Eden* by Thomas Bezucha (USA; 2003), and *The Business of Fancydancing* by Native American author and playwright Sherman Alexie (USA; 2002) have begun to explore different sets of meanings in which homosexuality and Amerindian ethnicities converge in unexpected ways. These rare occasions have not yet managed to change Euro-American perceptions of Native Americans bodies or their sexualities in any considerable way. The popularity of "American Indian" themes on the gay circuit is proof of this. "Cowboys and Indians," "chiefs," and "warriors" populate parties, nightclubs, gay parades and performances with steady regularity. Indians and cowboys maintain sex role connotations loosely based on cowboys' masculinity and the persistent feminization of the Native American in art and popular culture. Contrary to the United States and Canada, where gay, lesbian, and transgender Native people are becoming the sole legitimate users of indigenous attires and symbols, Native American-inspired hairstyles, clothing, and tattoos augment fashionable looks of gays around the world that are keen to spice up their style with exotic motifs derived from colonial scenarios.

Conclusions

Throughout history, Native Americans have been associated with homosexuality in a variety of different incarnations. From sodomy to degeneracy, through psychopathology and pornography, Native Americans have offered a language to express socially shared fears, anxieties, and desires, as well as un-conscious projections for numerous Euro-American people living in colonial and post-colonial contexts. This preliminary study has highlighted the significance that representations of sodomitic, feminized, depraved, or attractive Native American male bodies had in colonial history, and underlined the impact that such representations have on the present. Perhaps American Indian bodies appeal to gay Euro-Americans because many terms of deviance simultaneously converge in one symbol that successfully condenses the way in which most gay men see themselves. More generally, Euro-Americans' ambivalent fascination with the male Native American body, as it has been variably expressed through literature, art, and historical records, tied together in a seamless continuum unrelated, and often irreconcilable, ideologies framed by diametrical opposites of attraction and repulsion.

Although often noted at the margins of official history, the current examination of the ambivalent relationship with the Amerindian body enabled us to see that seemingly inconspicuous details that emerge from an analysis of cultural products such as paintings, narratives, and popular images contribute to hold together and reinforce ideological apparatuses based on unequal power relations between dominant and subordinate groups. I concluded this brief survey with an examination of gay men's popular take on Native American themes because perceptions and representations of Native Americans among them amply demonstrate that even discourses that circulate among contemporary subordinate minorities are, perhaps not surprisingly, couched on well-established tropes whose ideological underpinnings stem from colonial and imperialist practice and discourse. Although with the emergence of homosexual and gay subcultures Native America has offered models for acceptance of diversity, this benevolent, and maybe somewhat condescending, façade has also enabled a celebration of the vibrancy and exuberance of gay lifestyles through a convenient set of metaphors that dangerously resonate with paternalistic, or even racist, implications.

If a particular version of Native America still holds its allure for gay men despite bitter fights against ethnic stereotyping, and a vigorous championing of political correctness, it is because the meanings associated with it have become so intimately woven into the fabric of Euro-American consciousness that their expression can find no more suitable idiom than the one entrenched in the long established dichotomies of passive and active, looser and winner, the feminine and the masculine.

Today a handful of indigenous gays and lesbians are slowly starting to engage with issues of self-representation, and even some heterosexual artists are beginning to include homosexual, transgender, and lesbian themes in their art. Artists such as Kent Monkman (Cree/Métis), Virgil Ortiz (Cochiti Pueblo), Rose Bean Simpson (Santa Clara Pueblo), Waawaate Fobister (Ojibwa), and Hulleah Tsinhnahjinnie (Navajo/Creek/Muskogee), among others, have begun a process of redefinition of the visual terms in which Native American bodies can convey homoerotic, sexual, or queer messages that reach beyond the strict confines of Euro-American dominated gay networks (see Figure 14, photo gallery).[44]

Potentially, their work can have a considerable impact on both indigenous and larger public because they force audiences to imagine Native American bodies in unusual scenarios that open the possibilities not only for different readings and interpretations, but also for alternative discourses and practices that will inform realities yet to be imagined.

Notes

1. The term mostly used to describe the indigenous people of North America is "Native American." This term is often synonymous with "American Indian," which will also be used here. My use of the term "Amerindian" in addition to these other terms reflects colonial discourse that frequently referred with one term to all North, Central, Caribbean, and South American peoples. This term will be used in this sense.

2. See David F Greenberg, *The Construction of Homosexuality* (Chicago and London: University of Chicago Press, 1988); Jonathan Goldberg, *Sodometries: Renaissance Texts, Modern Sexualities* (Stanford: Stanford University Press, 1992).

3. See Jonathan Katz, *Gay American History: Lesbian and Gay Men in the USA* (New York: Thomas J. Crowell Company, 1976); Stuart Timmons, *The Trouble with Harry Hay: Founder of the Modern Gay Movement* (Boston: Alyson Publications, 1990); Evan B. Towle and Lynn M. Morgan (2002), "Romancing the Transgender Native: Rethinking the Use of the 'Third Gender' Concept," *GLQ: A Journal of Lesbian and Gay Studies* 8(4): 469–97.

4. Michel Foucault, *The History of Sexuality Vol. 1* (New York: Pantheon, 1978).

5. José Piedra, "Loving Columbus" in René Jara and Nicholas Spadaccini eds. *Amerindian Images and the Legacy of Columbus* (Minneapolis: University of Minnesota Press, 1992), 230–265.

6. Figure 12: Amerindians' sexual deviance and gender confusion appeared in early colonial iconography. In this detail of "The people and island that have been discovered . . ." by Johann Froschauer, likely to have been produced in Augsburg, Germany in 1505, two men of a different age, a mature bearded one and a beardless youngster, hint to both classical imagery of *erastes* and *eromenos*, as well as to biblical references of David and Jonathan, a male couple associated with platonic or homosexual love. Image by author is based on the original woodcut.

7. See Susi Colin, "The Wild Man and the Indian in Early 16th Century Book Illustration" in Christian Feest, ed., *Indians and Europe: An Interdisciplinary Collection of Essays* (Paris: Herodot, 1989), 26; Francisco Guerra, *The Pre-Columbian Mind: A Study Into the Aberrant Nature of Sexual Drives, Drugs Affecting Behaviour and the Attitude Towards Life and Death, With a Survey of Psychotherapy in Pre-Columbian America* (London and New York: Seminar Press, 1971); René Jara and Nicholas Spadaccini, eds., *Amerindian Images and the Legacy of Columbus* (Minneapolis and London: University of Minnesota Press, 1992). Previous analyses of this image have not considered this couple (for example, Hugh Honour, *The New Golden Land: European Images of America from the Discoveries to the Present Time* (New York: Pantheon Books, 1975), and *The European Vision of America* (Cleveland: The Cleveland Museum of Art, 1975); Robert F. Berkhofer, *The White Man's Indian: Images of the American Indian from Columbus to the Present* (New York: Vintage Books, 1979); Susi Colin, "The Wild Man and the Indian" (Paris: Herodot, 1989); Peter Mason, *Deconstructing America: Representations of the Other* (London: Routledge, 1990); and John Francis Moffit, *O Brave New People: The European Invention of the American Indian* (Albuquerque: University of New Mexico Press, 1988). The unspoken complicity between the two men, the intensity of their gesture, gaze, and body posture indicate early Renaissance conventions for describing male-to-male intimacy. David and Jonathan's embrace and gaze connoted this relationship (see James Smalls, *Gay Art* (n.p.) (New York: Parkstone Press, 2008). Secular scenes such as "Men's Bath-House" by Albrecht Dürer used this convention (see Giulia Bartrum, *Albrecht Dürer and his Legacy* (London: British Museum Press, 2002). It is tempting to suggest that the homoerotic allusions in this image may have been a visual reference for Froschauer's Amerindians.

8. In early Iberian colonial context the word "sodomy" was used to describe any non-canonical sexual practice including male-to-male sex (see Guerra, *The Pre-Columbian Mind*); Peter Sigal, *Infamous Desire: Male Homosexuality in Colonial Latin America* (Chicago: Chicago University Press, 2003); and José Piedra, "Loving Columbus" in René Jara and Nicholas Spadaccini, eds., *Amerindian Images and the*

Legacy of Columbus (Minneapolis: University of Minnesota Press, 1992), 230–265. Expressions such as "nefarious vice," "abominable sin," "unspeakable vice," and "acts against nature" were more often used with reference to male-male-sexual conduct. In this article I will use the term sodomy to mean male-to-male sexual liaisons as a convenient shortcut.

9. See Goldberg, *Sodometries,* 1992: 96–197; Piedra, "Loving Columbus," 1992: 261, note 25.

10. Theodore De Bry, *America de Bry, 1590–1634: Amerika oder die Neue Welt / die 'Entdeckung' eines Kontinents in 346 Kupferstichen, bearbeitet und herausgegeben von Gereon Sievernich* (Berlin: Casablanca, 1990), 177.

11. De Bry's "hermaphrodites" were not biologically intersexed individuals, but rather males donning women's clothes who performed women's tasks. See Katz, *Gay American History,* 1976; William Roscoe, *Changing Ones: Third and Fourth Genders in Native North America* (London: MacMillan, 1998), 12; and Walter Williams, *The Spirit and the Flesh: Sexual Diversity in American Indian Culture* (Boston: Beacon Press, 1986).

12. See Bernadette Bucher, *Icon and Conquest: A Structural Analysis of the Illustrations of De Bry's Great Voyages* (Chicago: University of Chicago Press, 1981).

13. The feminization of the Amerindian body continued well into the nineteenth century. In "America" (1864) by John Bell (British; 1811–1895); part of London Royal Albert Memorial's four corner allegories of the continents a man wearing a conventional feather skirt sits at the feet of an imposing female figure.

14. For example, Honour (1975), *The European Vision of America*: plates 5, 125, 245.

15. Honour (1975), *The European Vision of America.*

16. Antonello Gerbi, *La Disputa del Nuovo Mondo* (Milano: Adelphi, 2000 [1955]), 66, 68.

17. Michel Guillaume John de Crèvecœur, *Journey into Northern Pennsylvania and the State of New York* (Ann Arbor: University of Michigan Press, 1964 [1801]), 53.

18. See G. S. Rousseau and Roy Porter, eds., *Exoticism in the Enlightenment* (Manchester: Manchester University Press, 1990); Mason (1990), *Deconstructing America.* Joane Nagel pointed out that the feminization of the Other at the core of colonial and imperialist projects has parallels with Freudian theory about hysteria. In a recent paper she shows how the feminine failure of self-restraint is associated with deviance in dominant discourse (Nagel (2000), "Ethnicity and Sexuality," *Annual Review of Sociology* (26): 107–133). This correspondence has striking parallels with old theories about American peoples' intemperance and unrestrained sexual appetites highlighted in studies about the Other contained in early travelers descriptions. See Mason (1990), *Deconstructing America,* 56–57; Anthony Pagden, *The Fall of Natural Man: The American Indian and the Origins of Comparative Ethnology* (Cambridge: Cambridge University Press, 1982), 86.

19. Terence Turner, "The Social Skin" in Jeremy Charfas and Roger Lewin, eds., *Not Work Alone: A Cross-cultural View of Activities Superfluous to Survival* (Beverly Hills: Temple Smith, 1980), 112–140.

20. See Jean François Lafiteau, *Mœurs des sauvages américains comparée aux mœurs des premier temps* (Paris: Saugrain, 1724). Lack of masculinity and degenerate

virility are often co-terminus with homosexuality and effeminacy in colonial discourse. Interestingly, during the Enlightenment the French used the term *berdache* to describe the effeminate, receptive male. This was the same term they used to describe Native American transgender individuals: see Greenberg (1988), *The Construction of Homosexuality*: 333–334, note 204.

21. Stephanie Pratt, *American Indians in British Art: 1700–1840* (Norman: University of Oklahoma Press, 2005), plate 10.

22. Honour (1975), *The European Vision of America*: plate 281; Véronique Wiesinger, ed., *Sur le Sentier de La Découverte: Rencontres Franco-Indiennes du XVIe au XXe Siècle* (Paris: Editions de la Réunion des Musées Nationaux, 1992), 72–75.

23. Greenberg (1988), *The Construction of Homosexuality*.

24. George Devereux (1937), "Homosexuality among the Mohave Indians," *Human Biology* 9: 498–527; Willard W. Hill (1935), "The Status of the Hermaphrodite and Transvestite in Navaho Culture," *American Anthropologist* 37: 273–9; Alfred Kroeber (1940), "Psychosis and Social Sanction," *Character and Personality* 8: 204–15; Elsie Clews Parsons (1939), "The Last Zuni Transvestite," *American Anthropologist* 41(2): 338–40.

25. Joyce E. Chaplin, *Subject Matter: Technology, the Body, and Science on the Anglo-American Frontier, 1500–1676* (Cambridge: Harvard University Press, 2001); Piedra (1992), "Loving Columbus."

26. Pratt (2005), *American Indians in British Art*, 23.

27. George Catlin, *Letters and Notes on the Manners, Customs, and Conditions of the North American Indians Written During Eight Years' Travels (1832–1839) amongst the Wildest Tribes of Indians of North America* (New York: Dover, 1973 [1844]) vol. II, 123. Parallels drawn between Indian bodies and classical aesthetics are common in European-American literature. Recent textual analysis of seminal texts and frontier letters has offered irrefutable evidence that homoerotic meanings permeate diaries and fiction describing both imagined and real-life encounters and relationships between Euro-Americans and American Indians: see William Benemann, *Male-Male Intimacy in Early America: Beyond Romantic Friendship* (Binghamton, NY: Haworth Press, 2006); Chris Packard, *Queer Cowboys and Other Erotic Male Friendships on Nineteenth-century American Literature* (New York: Palgrave, 2006); *Men in Eden: William Drummond Stewart and Same-Sex Desire in the Rocky Mountain Fur Trade* by William Benemann (Lincoln: University of Nebraska Press, 2012).

28. Tim Fulford, *Romantic Indians: Native Americans, British Literature, and Transatlantic Culture, 1756–1830* (Oxford: Oxford University Press, 2006), 146.

29. Leslie A. Fiedler, *Love and Death in the American Novel* (London: Penguin, 1966), 182.

30. Fiedler (1966), *Love and Death in the American Novel*, 214.

31. Rousseau and Porter (1990), *Exoticism in the Enlightenment*, 5–7.

32. Peter Van Lent, "'Her Beautiful Savage': the Current Sexual Image of the Native American Male' in S. Elizabeth Bird, ed., *Dressing in Feathers: The Construction of the Indian in American Popular Culture* (Boulder, CO: Westview Press, 1996), 211–244.

33. Kobena Mercer, "Skin Head Sex Thing: Racial Difference and the Homoerotic Imaginary" in Coco Fusco and Brian Wallis, eds., *Only Skin Deep: Changing Visions of the American Self* (New York: Harry Abrams Inc. Publishers, 2003), 237–265.

34. David Jenkins (1993), "The Visual Domination of the American Indian: Photography, Anthropology, and Popular Culture in the late Nineteenth Century," *Museum Anthropology* 17(1): 9–21; Aleta Ringlero, "Prairie Pinups: Reconsidering Historic Portraits of American Indian Women" in Coco Fusco and Brian Wallis, eds., *Only Skin Deep: Changing Visions of the American Self* (New York: Harry Abrams Inc. Publishers, 2003), 183–197.

35. Helen Carr, "Woman/Indian, the American and its Others" in Peter Hulme and Francis Barker, eds., *Europe and its Others Proceedings of the Essex Conference on the Sociology of Literature 2* (Colchester: University of Essex, 1984), 46–60; Louis Montrose, "The Work of Gender in the Discourse of Discovery" in Stephen Greenblatt, ed., *New World Encounters* (Berkeley: University of California Press, 1993), 117–217; Nagel (2000), "Ethnicity and Sexuality"; Rousseau and Porter (1990), *Exoticism in the Enlightenment.*

36. Harry Bush, *Hard Boys* (San Francisco: Green Candy Press, 2007).

37. Scott Simmon, *The Invention of the Western Film* (Cambridge University Press, 2003), 23–24.

38. Elizabeth Cromley (1996), "Masculine/Indian," *Winterthur Portfolio* 31(4): 265–80; Philip J. Deloria, *Playing Indian* (New Haven and London: Yale University Press, 1998).

39. George Quaintance, *The Art of George Quaintance* (Berlin: Janssen Verlag, 2003).

40. See Walter Kundzicz, *Champion* (New York: Goliath Books, 2004); Bob Mizer (2007), *Athletic Model Guild (AMG): American Photography of the Male Nude 1940–1970*, vol. 7 (n.p.) Janssen; Physique Pictorial, *The Complete Reprint of Physique Pictorial: 1951–1990.* 3 vols. (Köln: Taschen, 1997).

41. Harry Hays's cohort included people that became renowned anthropologists active in research about *berdaches* and Two-Spirit people (collective name for Native American lesbians, gays, bisexuals and transgenders). These are Sue-Ellen Jacobs, William Roscoe, and Walter Williams (see Timmons (1990), *The Trouble with Harry Hay*).

42. Past and recent production of gay cartoons, comic strips and vignettes depicting Native Americans with erotic and pornographic slant are too many to mention. Twentieth-century porn comics such as the series *Rough Trade,* for example, featured Native American scenarios. Virtually every gay illustrator since the beginning of this art genre has created images of naked Native Americans, or imagined sexual narratives in which they appear. Typically this can be seen in the art of illustrators that go under the names of Adrien, Etienne, Mike, Dupre, Joe Philips, Wolf, Harry Bush, David Kawena, Steven Mochica, and graphic studios such as Krayel Studio, Kosen Studio, DC Vertigo, and many others.

43. Soft porn products such as calendars, fine art prints and amateur photography show a number of Native American models or themes (e.g., Dan Nelson, Studio Morphosis), but pornographic magazines featuring Native American themes are rare. Several porn models that are marketed as "Latino" or "Hispanic" display clear indigenous phenotypes but are never presented as American Indian. Native American adult actors rarely appear as themselves, an exception is the model known as Daniel in a spread entitled "Hung like Crazy Horse" in the gay pornographic magazine *BlueBoy* (October 1997).

44. See Aleta Ringlero (2006), "The Virtuoso," *American Indian NMAI* 7(4): 27–36; William Roscoe, *Changing Ones: Third and Fourth Genders in Native North America* (London: MacMillan, 1998); Glenn Sumi (2008), "Fobister's a Phenom: Waawaate Fobister Brings Down the House at Solo Buddies Opener," *Now* 28(5) at http://www.nowtoronto.com/stage/story.cfm?content=165173 [accessed 2 August 2009]; Kerry Swanson (2005), "The Noble Savage was a Drag Queen: Hybridity and Transformation in Kent Monkman's Performance and Visual Interventions," *Emisférica: Performance and Politics in the Americas* 2(2): 1–19.

6

Devil with the Face of an Angel

Physical and Moral Descriptions of Aboriginal People by Missionary Émile Petitot

MURIELLE NAGY

The French Oblate missionary, Émile Petitot (1838–1916), who lived in the Canadian northwest from 1862 to 1881, wrote about the language, traditions, history, and territory of the Dene[1] and Tchiglit (Siglit) Inuit.[2] Petitot was a prolific author.[3] However, with the exception of his contribution to geography (which included maps with Aboriginal toponyms and ethnonyms[4]) and his dictionaries (which have been used by linguists[5]), his ethnographic texts and travel narratives were almost forgotten, until they were inventoried and indexed with thematic excerpts in the early 1970s.[6] This chapter will be interrogating these texts.[7]

Petitot wrote for various audiences: his religious order; readers of journals that promoted Catholic missions; the scientific community; and, in the later part of his life, the general public. Except for frequent statistics on the Aboriginal people he baptized, Petitot's writings are not specifically concerned with his missionary endeavor. Petitot was first and foremost an avid researcher in ethnography, linguistics, and geography—and was considered as such by his Oblates colleagues and superiors. He positioned himself as a detached observer and would certainly have claimed that he wrote in what Mary Louise Pratt has called an "innocent pursuit of knowledge."[8] Pratt's expression is linked to her definition of "anti-conquest," as being "the strategies of representation whereby European bourgeois subjects seek to secure their innocence in the same moment as they assert European hegemony."[9] Petitot certainly did not view himself as an instrument of "European hegemony," and indeed probably saw himself more as an explorer and ethnographer than a missionary.

85

As a consequence, Petitot's writings can be read and utilized as primary sources for the period of early contact between Aboriginal peoples and missionaries, in what is now the Northwest Territories of Canada. Many of his comments on the different peoples he lived with show an openness toward their cultural practices that was unusual for someone working as a missionary. An acute observer, Petitot recorded facial expressions and body language, as well as describing and drawing in great detail body features, decorations (earrings, labrets, tattoos, etc.), and hair styles. Notwithstanding his finely tuned cultural observations, comments about his own values, emotions, and sexuality are also embedded throughout his writings on Aboriginal peoples, and the literary tropes he uses position the indigenous body as a site of erotic fascination.

This chapter will explore how Petitot's physical and moral descriptions of Aboriginal individuals are full of contradictions and require a careful and nuanced reading. His own correspondence and that of others about him both reveal paradoxes between the missionary and the man. The letters show that all the significant aspects of Petitot's life in Canada were linked to his intimate relationships with Aboriginal people, and that these influenced—whether consciously or not—what he wrote in his publications.

Since Petitot wrote within a colonial context, this analysis will be organized around three specific considerations. First, it follows Edward Said's recommendations for the study of Orientalist texts: "The things to look at are the style, figures of speech, setting, narrative devices, historical and social circumstances, *not* the correctness of the representation nor its fidelity to some great original."[10] It also engages with Said's general views on Orientalism—a school of interpretation and representation of the Orient from a Western perspective—to reflect on the notion of otherness in Petitot's writings. As a missionary, Petitot was very much part of the colonial invasion of the Canadian North and his depictions of Aboriginal peoples were addressed to European readers, for whom they were totally foreign and exotic.

Second, this analysis considers the "relationship between personal narrative and impersonal description in ethnographic writing" that has been both identified and studied by Mary Louise Pratt.[11] Petitot mostly traveled with guides and often lived with the families he visited. During those times, he had little privacy and his writings provide many details about the people he met, the social settings he encountered, and even the conversations he heard. Petitot was the narrator in his books as well as a major protagonist in the situations he described. His status as an agent of religious change, his preconceptions about the Aboriginal people he met, and the interactions he had with them, all influenced his depictions of various individuals and groups. Focusing on the intimate will allow examination of the "strangely familiar that proximities and inequalities may produce" in colonial social relations.[12]

Third, this analysis explores how several characteristics of Petitot's correspondence illustrate Ann Laura Stoler's observations about "archives as condensed sites of epistemological and political anxiety rather than skewed and biased sources."[13] Under consideration here will be how Petitot often mentioned one specific body alteration—circumcision—as being common among the Dene, although his colleagues thought it was not practiced, and also Petitot's increasing fear of being killed by Aboriginal people during his long stay in the North, when this threat was discredited by his fellow Oblates. As Stoler explains, "imagining what *might be* was as important as knowing what was,"[14] and this was certainly the case with Petitot's beliefs about the various Aboriginal people he encountered.

Petitot's publications, his letters, and those of others about him,[15] have to be read and understood within the context of the Oblates' epistolary style. In their correspondence, a deferential tone was used by the Oblate brothers and fathers toward their superiors, who often addressed them as "my child." The style of the letters is usually flowery, but also confusing at times due to the inclusion of ambivalent terms or euphemisms which understate the sordid nature of the sexual behavior of some Oblates. Furthermore, the way the Oblates infantilized Aboriginal people when writing about them did not always give an accurate idea of their real age. Petitot wrote about "little children of 10 to 12 years old," a "young man" who was 28 years old,[16] and another who was much younger, being 14 years old.[17] Terms such as "young boy" and "young child" were often given to boys between 10 to 15 years old working for the Oblates as servants and guides. At 16 years old, they were hired as *engagés* and entitled to a raise in salary.[18] An 1875 letter from Bishop Grandin is quite revealing about the Oblates' way of assessing adulthood among Aboriginal people, as he was asking the Canadian government to make "a law giving paternal authority to the Oblates on the Savage children, up to 20 to 22 years old for the boys and 15 or 18 years old for the girls."[19]

Petitot was among the first scholars to try to demonstrate the Asiatic origin of the Inuit and the Dene from his study of their languages and oral traditions.[20] He believed in a common origin for all people, but contrary to evolutionist theories of his time, which explained the development of human cultures from "savage" to "civilized,"[21] he thought that some groups had degenerated.[22] The theory of degeneration goes back to the Middle Ages and the Christian view that all humans were from the Garden of Eden (located in the Near East) and shared God's revelation of himself and his wishes, while some groups moved away and "degenerated" into polytheism, idolatry, and immorality.[23] That was how Petitot explained that attributes belonging to various periods of evolutionism could be found together among the Dene who were, he claimed, the descendants of one of the ten lost tribes of Israel.[24] Here, Petitot's understanding was close to the thinking of eighteenth-century

Jesuit missionary Lafitau[25] who argued that the customs and religions of Amerindians resembled those of ancient Greeks and Romans since they were altered versions of a common history.

Petitot's adherence to the Jewish Indian theory was also in line with the writings of seventeenth-century British writers such as Thomas Thorowgood, who used it to place the Amerindians into the Biblical narrative, thereby rendering them less foreign and threatening.[26] As Julia Cohen writes, texts about the "Lost Tribe" theory "initiate a system of otherness, using rhetorical techniques of simultaneous appropriation and disassociation with the other." She adds that for the English Puritans, "speculating about both the identities of the Native Americans and the Jews, and creating a type of amalgamated identity of both cultures in the body of the Amerindian, was a way of diverting fears of otherness."[27] This process of forcing a Jewish identity onto the Dene is explicit in Petitot's writings: he continuously referred to similarities between the Jewish and Dene traditions, as well as drawing some dubious parallels between Hebrew and Dene languages.[28]

In order to demonstrate that the Dene were descendants of the tribe of Dan, Petitot insisted that many Dene groups practiced circumcision. He wrote that a former Hare female shaman and an old Dinjié (Gwich'in) female chief told him that their people "circumcised their male children a few days after they were born, with a piece of flint" and that "circumcision was also practiced for skin diseases."[29] Petitot indicated that the Montagnais (Chipewyan) and the Dogrib did not practice circumcision, but that "it seems that the Indians of the Rocky Mountains observe it faithfully."[30] After learning this fact, Petitot read that Alexander Mackenzie "thought he saw traces of circumcision among the Hare" during his 1789 travel.[31]

Petitot explained that: ". . . circumcision is honoured . . . ; if practiced in almost all tribes, it is not the case for all families and for each individual. Until the past years, some were not circumcised."[32] In a later publication, he wrote: "Some Dinjié told me that an adult that was not circumcised after his birth would need to accomplish this operation himself without the help of someone else."[33] While at Fort Anderson in November 1865, where people were dying of scarlet fever, he saw a "young [Gwich'in] boy who had been circumcised in haste when the epidemic started since the ceremony had been deferred year after year due to negligence."[34] Petitot even noted in his books whether individuals he met were circumcised[35] and reported a conversation between two Hare women in which they disparaged white men for being uncircumcised.[36] However, the Slavey Petitot met were not circumcised and called "dogs" uncircumcised men, "a friendly and unshameful epithet."[37] As for the Tchiglit Inuit, Petitot stated that they adopted circumcision in the 1870s, following the example of the Gwich'in. An Inuk living among the Gwich'in and married to one of them asked Petitot to circumcise him in 1876, as the man felt rejected by the Gwich'in for not having it done yet.[38]

Much interested in circumcision, and well versed about it, Petitot compiled a list of other cultures where it was performed and at what age.[39] However, with the exception of Mackenzie's comment, Petitot's observations concerning circumcision have never been corroborated, and in fact were harshly criticized by Father Ducot (1848–1916)[40] who lived among the Dene beginning in 1874.[41] Thus, his reports of circumcision among the Dene may seem initially surprising, if not erroneous. As Father Morice (1858–1938) wrote, Petitot had pushed his habit of assimilation so far that he might have exaggerated his conclusions to benefit his thesis of the Jewish origins of the Dene.[42] His fixation on circumcision is nevertheless more comprehensible if one knows the context of his private life during his long stay in Canada. Through the "ethnographic gaze," Petitot's repeated allusions to circumcision might have allowed him "to focus on a world of virility and masculinity without revealing an erotic fascination."[43]

Petitot started to publish about Aboriginal peoples of the Canadian Northwest as far back as the 1860s, but it was only after his return to France in 1883 that he wrote books about his travels aimed at the general public. In his introductions, Petitot insisted that his travel narratives were truthful and warned about the morals of the peoples he wrote about. However, wanting his readers to appreciate Aboriginal peoples, he stated, for example: "And the Dene are called savages! Come on! It would be a joke or a deception if not at the same time a gross and undeserved insult. With regard to their intelligence, honesty, and qualities of the heart, these Dene are not inferior to any farming population. I even think that they are superior to the latter. . . ."[44] He was also open to interpreting the moral codes of his hosts through their eyes as in this comment about begging:

> . . . among the Red-Skins, . . . begging does not indicate misery but superiority. The one who asks and receives freely, deducts a right, a tribute from foreigners, subordinates. The more the Indian is proud and haughty, the more he begs. The Chipewyan and the Eskimos are a lot like that. The Dogrib, even more. On the contrary the Slave, Hare, and Loucheux [Gwich'in] never ask anything for free because they have more humility.[45]

Thus, Petitot applied character traits to specific groups, and implicitly associated the most humble with those who were Christianized. Hence, while relativizing parts of the Aboriginal cultures[46] and genuinely trying to open the mind of his readers, Petitot still created distance between them and the

Other. Whether conscious or not, as Stoler noted, "The language of difference conjured up the supposed moral bankruptcy of culturally dissonant populations, distinguishing them from the interests of those who ruled."[47] Indeed, when depicting Aboriginal peoples, at times Petitot used a denigrating style starting with praise and ending with derogatory comments, as in the following examples:

> . . . I am convinced that it is the savage adornment of the Eskimos that makes all their glory, all their prestige, all the reputation of bravery they have among their neighbours and even in the mind of the Europeans. Take away from this inhabitant of the Polar ices those long magical fringes of animal furs that circle by levels his arms and his legs, cut this red mane of wolverine which bristles up around his sardonic face, strip him from his talismans, . . . dress him humbly with the gloomy and modern clothes of the civilized man, and as soon you will take away all the audacious and brave character, you will only see a timid being, ugly, clumsy and ashamed, despising himself, and no doubt despicable.[48]

> . . . note this [Eskimo] exquisite politeness, this inventive and imitating genius, this aptitude for arts and trade; observe mainly this insolence, this lack of fear, this flaw of shame and honesty.[49]

Comparing Inuit and Indians, Petitot used the same stylistic tactic, complimenting the former but criticizing the latter, as in these examples: ". . . there was nothing in the manner of these women that resembled the Red-Skin coarseness."[50] "The Eskimo is not timid, taciturn like the Red-Skin. He is confident, talkative. . . ."[51] "[The Eskimos] are lazy by egoism, not by nature; while the Red-Skin is so fundamentally lazy that he will prefer to miss the essential than to work."[52] "Anger is certainly the dominant passion of the Eskimos; but it is due to their sensitivity, which is more developed than among the Red-Skins, and accompanied by a bigger dose of pride."[53]

In another part of his book on the Inuit, incorporating the "savage as child" analogy,[54] Petitot reversed his dichotomic trope by starting with an infantilizing comment followed by praising ones: "They talk with naivety and intelligence. Their child-like minds are curious and eager to learn."[55] Elsewhere, Petitot described his first impression of the Yellowknife Dene in the following terms: "Aside from the religious question, they were big and pleasant children. As such, they had their simplicity without their mischief."[56] Even when trying to be complimentary, Petitot ended with a condescending tone: "There is nobody in the world who loves to talk more about sciences and arts than the Savage peoples of America. They are restless for instruction and have the curiosity of intelligent and sensible children."[57] In all these examples, Petitot was following the colonial discourse of his time, in which "racialized Others invariably have been compared and equated with children, a representation that conveniently provided a moral justification for imperial policies of tutelage, discipline and specific paternalistic and maternalistic strategies of custodial control."[58]

Petitot also used his demeaning technique to depict individuals. For example, when describing four Inuit men, he first made short physical depictions of their faces with emphasis on the shape of the nose and the eyes. Then, using a hue continuum, he qualified the man with a white skin as good-looking; the one with the pink complexion "of an European" as having a serious look; but those with brown skins as looking respectively treacherous and false "like a Maltese sailor," and insolent and rebellious.[59] About an Inuk teenager, he noted that "His walk was not dawdling like that of the Red-Skin, . . . but easy, noble, brisk like that of a European."[60] As for the Métis, they are good-looking "only when the proportion of white blood dominates that of red blood."[61] Elsewhere, describing one of his young Gwich'in guides, he stated, "His face was gentle and pleasing, but with uninteresting placidity and no passions. His regular and agreeable features were void of expression."[62] Later, about his same guide's two sisters: "Pretty dolls but stupid, with an empty gaze, an indifferent and frozen face."[63]

In all the examples presented above, the question arises why Petitot was so negative about the people he lived with and ministered to. Clues to the answer can be found in one of Petitot's most elaborate attempts at playing the physiognomist while depicting his Dene guide, Hyacinthe Dzan-yu,[64] whom he considered "an excellent specimen" of a Dene adolescent:

> Beautiful type of Dene this one, half-Slavey half-Hare, an Adonis among his compatriots, which are not the most beautiful: big black eyes, intelligent and soft, shadowed by long eyelashes, strong and high eyebrows, strait nose flat in the middle, lips disdainfully curled up, high forehead but narrow. However, this nice physiognomy sometimes showed cunning signs. His gaze usually jovial and friendly would become perfidious. His stretched neck changed into something abject, his narrow temples would tighten with stubbornness. There was then a *je ne sais quoi* of the devil in this face of an Indian angel.[65]

Here again, Petitot alternates between compliments and criticisms, and ends his depiction with a vision of the devil lurking beneath the surface. Petitot had previously used the same paradoxical analogies about humans in general, writing that "once one is exalted, one is soon despised; a double being, participant to the nature of the angel and the demon!"[66] However, and as will be seen, his comments did not originate from his views toward others, nor the Other, but had more personal roots.

Further clues to the source of Petitot's attitudes toward the Dene and Inuit can be found in his moral portrait of Dzan-yu who had admitted having "known the passions of full grown men" at a young age and depicted himself as vicious. Yet, Petitot insisted that he was a very honest and excellent Christian who became the most educated person of his tribe, having learned how to read, write, do arithmetic and sing. Petitot even added that he was faithful to his wife.[67] Archival letters actually revealed that Dzan-yu's marriage took place in the fall of 1866 and was very likely arranged to dissipate any gossip about Dzan-yu and Petitot.[68] Indeed, shortly after arriving in Fort Good Hope in 1865, Petitot became infatuated with Dzan-yu, who was then 16 years old.[69] Although it is not clear when and from whom Dzan-yu received his Christian name, he was likely baptized by Petitot who must have known that in Greek mythology Hyacinthe refers to a beautiful youth beloved by the gods Apollo and Zephyr.

Petitot's private relationship with Hyacinthe[70] casts a different light on his description of him. As Shumaker remarks, "who is labelled abject varies according to the prejudices of those who project the abject onto the 'other' to suppress recognition of it inside themselves."[71] Furthermore, within the colonial context, "the negative vision" of Aboriginal people "is important to the colonizer's identity because it provides him with an 'imaginary' Other onto whom his anxieties and fears are projected."[72] According to his letters to his Oblate superiors, Petitot felt very guilty about his relationship with Hyacinthe. Although Stoler warns that "power shapes the production of sentiments and vice versa,"[73] there is no evidence in the letters to suggest that Petitot forced Hyacinthe into the relationship. While it is impossible to determine whether there was any overt abuse or coercion in Petitot's relationship with Hyacinthe, he was nonetheless breaking his vows of celibacy.

Petitot reported that Hyacinthe despised him after his public confession of their relationship, and that "irony and sarcasm had replaced affection."[74] However, Hyacinthe's refusal to leave Petitot—he was his guide on and off for over ten years—suggests that their attachment was more nuanced than Petitot as the abuser and Hyacinthe as the victim. Since power is exercised in the game of unequal relationships,[75] evidently a complex relation of dependence and power was established between the two men. Hyacinthe's source of power was his knowledge of the territory and how to live in it, as well as being an amorous and erotic attraction for Petitot. It is hard to tell if this attraction was reciprocal for Hyacinthe but by hiring him as his guide, Petitot gave him a status, wages, and access to European material goods which were rare at that time. Yet for Petitot, the relationship was profoundly troubling, and the guilt that haunted him is implicit in his depiction of Hyacinthe as an object of "fear and desire."[76]

According to Petitot, from 1868 his health deteriorated "physically and morally because of sarcasms and calumnies from Dene towards him."[77] His correspondence and that about him refer to his relationships with other young Dene men,[78] his paranoiac and mystic crises, as well as his obsession with circumcision. Petitot was convinced that the Dene wanted to kill him, and that the only way to appease them was by getting circumcised, which he attempted a few times.[79] This extreme action might reflect his fantasy to be like the imaginary Other (the circumcised Dene), but also a way to mutilate himself to restrain his desire. Although probably linked to his anticipation of punishment for his sexual misconduct, Petitot's fear of being killed by Aboriginal people was deemed unreasonable by the Oblates living with him.[80] Nevertheless, his anxiety is also understandable in a colonial context, where "the portent-laden future of revolt and betrayal is always on the imminent and dangerous horizon."[81] Indeed, even in his books, he believed the threat of being killed by the people he encountered was very present since some Inuit had already threatened to do it.[82]

The correspondence of the Oblates about Petitot indicates that they thought he had lost contact with reality, and experienced hallucinations and delusions—all signs of a psychosis. However, his delirium of persecution and his fixation on self-circumcision evaporated as soon as he left the North. Hence, during his two-year sojourn in France (1875–1876), and after his definitive return to France in 1883, he does not seem to have had other crises. Petitot appears to have been among "those Europeans confused by their own inappropriate desire and disdain."[83] His mental state was affected by his forbidden and, to some degree, frustrated love for Hyacinthe and probably led him to live in delusion about their relationship. His public confession brought guilt, ambiguity, as well as public and private pressure regarding not only the relationship but also his vows to the Church. In addition, he suffered long periods of isolation at the missions when Aboriginal people went back to their camps which, he admitted, depressed him.[84] Even when in the company of people, the fact that he was a lone European missionary living for an extended period in a very different context from his own, plausibly provoked profound culture shock.

In his 1893 book, Petitot wrote that he finally regained his health and "spent happy days" at Cold Lake (Alberta) at the end of his sojourn in Canada. The reality was, however, different, since many letters about him attest that he had periods of distress and acted impulsively.[85] Although he did not publish about it, he fell in love with Nikamous, a Métis woman whom he married (*à la façon du pays*) after trying to quit the Oblates order[86] and lived with for a few months in 1881.[87] However, the Oblates did not allow Petitot to live a layman's life; they kidnapped him and sent him to an asylum in Montreal in 1882.[88] Back in France the next year, he left the Oblates and became a secular priest. Writing allowed him to return to his fantasies of the Other and, to some degree, to voice his frustrations at the Oblates. Without naming them, he denounced how his reputation had been defamed and his freedom taken away from him.[89] However, and probably since at least two[90] of his five books aimed at a general audience were published by Catholic publishers promoting missionary work, Petitot maintained a moralizing tone where a "rhetoric of inequality . . . tolerates all manner of contradiction."[91] The tension between Petitot's roles as an ethnographer and a missionary may have prompted him to pepper his writings with Biblical references and phrases reflecting the colonialist views of the nineteenth century.

In Petitot's books for the general public, his biases and personal demons are evident. Amidst his ethnographic and geographic descriptions, Petitot went back and forth into his missionary mode, including theories influenced by his understanding of the Bible and the imperial racism of his time. His moral and physical descriptions of Aboriginal people were as ambivalent as he probably was about himself. Personal thoughts often tainted his writings which held two discourses, often intertwined: one colonialist and racist, and in contrast, the other personal and sentimental. As Stoler so eloquently puts it, "How and to whom sentiments of remorse or rage, compassion or contempt were conveyed and displayed measured degrees of social license that colonial relations inequitably conferred."[92] Indeed, Petitot's trope of alternating between the good and the bad—the angel and the devil—reflected as much about himself as what he thought of the Other.

Acronyms

AASB = Archives of the Archdiocese of Saint Boniface, Winnipeg
AD = Archives Deschâtelets in Ottawa
AOR = Archives of the Oblates in Rome
PAA = Provincial Archives of Alberta
RCAY = Roman Catholic Archives in Yellowknife

Notes

1. The Dene are a subarctic people speaking Athapaskan languages. The main Dene subgroups that Petitot visited were the Chipewyan, Dogrib, Gwich'in, Hare, Slavey, and Yellowknife. See Émile Petitot, *Monographie des Dènè-Dindjié* (E. Leroux éditeur: Paris, 1876); *Dictionnaire de la langue Dènè-Dindjié* (E. Leroux éditeur: Paris, A. L.

Bancroft and Co.: San Francisco, 1876); *Traditions Indiennes du Canada Nord-Ouest* (Littératures populaires, Tome XXIII, Maisonneuve frères et Ch. Leclerc: Paris, 1886).

2. Since the 1980s, the Inuit of the western Canadian Arctic call themselves Inuvialuit. See Émile Petitot, *Monographie des Esquimaux Tchiglit* (E. Leroux éditeur: Paris, 1876); *Vocabulaire français-esquimau*: (Bibliothèque de Linguistique et d'Ethnographie américaines publiée par A. L. Pinart, vol. 3, E. Leroux éditeur: Paris, A. L. Bancroft and Co.: San Francisco, 1876).

3. See Émile Petitot *Les Grands Esquimaux* (E. Plon, Nourrit et Cie: Paris, 1887); *En Route pour la Mer Glaciale* (Letouzey et Ané: Paris, 1888); *Quinze Ans sous le Cercle Polaire,* Tome I, Mackenzie, Anderson et Youkon (E. Dentu: Paris, 1889); *Autour du Grand Lac des Esclaves* (Nouvelle Librarie Parisienne, Albert Savine: Paris, 1891); *Exploration de la région du Grand Lac des Ours* (Téqui: Paris, 1893).

4. See Donat Savoie (ed.), *Land occupancy by the Amerindians of the Canadian Northwest in the 19th century as reported by Émile Petitot. Toponymic inventory, data analyses, legal implications* (Indian and Northern Affairs, Canada: Ottawa; Occasional Publication, 49, CCI Press: Edmonton, 2001).

5. See Ronald Lowe, *Les trois dialectes inuit de l'Arctique canadien de l'Ouest. Analyse descriptive et étude comparative* (Collection Travaux de recherche, 11, GÉTIC: Sainte-Foy, 1991).

6. See Donat Savoie, *Les Amérindiens du Nord-Ouest canadien au 19e siècle selon Émile Petitot* (Travaux de recherche sur le delta du Mackenzie, Ministère des Affaires Indiennes et du Nord Canadien: Ottawa, 1971).

7. Funding for my research on Petitot was obtained through a Social Sciences and Humanities Research Council (SSHRC) research grant, and posdoctoral fellowships from SSHRC, the Fonds pour la formation de chercheurs et l'aide à la recherche (FCAR), the Faculté des Sciences Sociales, and the CÉLAT of Université Laval.

8. Mary Louise Pratt, *Imperial Eyes. Travel Writing and Transculturation* (Routledge: London, 1992), 84.

9. Pratt, *Imperial Eyes*, 7.

10. Edward Said, *Orientalism: Western Conceptions of the Orient* (Penguin: London, 1994 [1978]), 21, emphasis in the original.

11. Pratt, *Imperial Eyes*, 27.

12. See Ann Laura Stoler "Tense and tender ties: The politics of comparison in North American history and (post) colonial studies" in *Haunted by Empire: Geographies of Intimacy in North American History*, ed. Ann Laura Stoler (Duke University Press: Durham, 2006), 14.

13. Ann Laura Stoler, *Along the Archival Grain. Epistemic Anxieties and Colonial Common Sense* (Princeton University Press: Princeton, 2009), 20.

14. Stoler, *Along the Archival Grain*, 21, emphasis in the original.

15. Although Savoie (*Les Amérindiens*) identified most of Petitot's correspondence, I found additional letters about him in various archives.

16. Petitot, *Les Grands Esquimaux*, 101, 152.

17. Petitot, *Exploration*, 364, 369.

18. Grandin to Taché, Caribou Lake, November 28, 1866 (PAA).

19. Grandin to the Ministry of the Interior in Ottawa, April 5, 1875 (PAA), my translation from French.

20. See Émile Petitot, "Observations du R. Père Petitot," *Congrès International des Américanistes, Nancy 1875, Compte-rendu de la première session*, tome 1 (Maisonneuve et Cie: Paris, 1875), 141–142; "De l'immigration asiatique," Congrès International des Américanistes, Compte-rendu de la première session, tome 2 (Maisonneuve et Cie: Paris, 1875), 245–256; *Origine asiatique des Esquimaux. Nouvelle étude ethnographique* (Espérance Cagniard: Rouen, 1890).

21. E.g., Lewis Henry Morgan, *Ancient Society* (Holt: New York, 1877); Edward Burnett Tylor, *Primitive Culture* (J. Murray: London, 1871).

22. See Émile Petitot, *Accord des mythologies dans la cosmogonie des Danites arctiques* (Émile Bouillon: Paris, 1890).

23. Bruce Trigger, *A History of Archaeological Thought* (Cambridge University: Cambridge, 1989), 33–34.

24. See Gilles Cadrin, "Le Père Émile Petitot et l'origine des peuples d'Amérique: Polygénisme ou monogénisme," *Francophonies d'Amérique* 2 (1992): 139–149.

25. Joseph-François Lafitau, *Mœurs des sauvages américains comparées aux moeurs des premiers temps* (La Découverte: Paris, 1994 [1724]).

26. Amy H. Sturgis, "Prophesies and Politics: Millenarians, Rabbis and the Jewish Indian Theory," *The Seventeenth Century* 14(1) (1999): 17.

27. Julia Cohen, "Members of the Tribe: Jewish-Amerindian Theory and the Making of a Modern American Consciousness," *COPAS*, 7 (2006) online at www-copas.uniregensburg.de/articles/issue_7/Julia_Cohen.php.

28. E.g., Petitot *Accord des mythologies*.

29. Émile Petitot, *Monographie des Dènè-Dindjié*, 77–78. I translated all the quotes from Petitot from French.

30. Petitot, *Monographie des Dènè-Dindjié*, 78.

31. Alexander Mackenzie, *A Journey from Montreal to the Polar and Pacific Oceans* (The Narrative Press: Santa Barbara, 2001 [1801]); Petitot, *Monographie des Dènè-Dindjié*: 78.

32. Émile Petitot, "L'Origine Asiatique des Indiens de l'Amérique du Nord," *Les Missions Catholiques de Lyon*, t. 11(546) (1879): 566.

33. Petitot, *Quinze Ans sous le Cercle Polaire*, 311.

34. Ibid., 177.

35. E.g., Petitot, *Quinze Ans sous le Cercle Polaire*, 218.

36. Ibid., 228, 230.

37. Petitot, *Autour du Grand Lac des Esclaves*, 347.

38. Petitot, *Les Grands Esquimaux*, 293.

39. Petitot, *Quinze Ans sous le Cercle Polaire*, 183.

40. Father George Ducot's criticisms were annotations he wrote in 1903 directly on the margins of a copy of Petitot's *Quinze Ans sous le Cercle Polaire* that I consulted at the Roman Catholic Archives of Yellowknife.

41. Gaston Carrière, *Dictionnaire biographique des Oblats de Marie-Immaculée au Canada*, tome 1 (Éditions de l'Université d'Ottawa: Ottawa, 1976), 311.

42. Adrien Gabriel Morice, "L'abbé Émile Petitot et les découvertes géographiques du Canada. Étude géographico-historique," *Bulletin de la Société Neuchâteloise de Géographie*, 29 (1920): 11–12.

43. Rudi C. Bleys, *The Geography of Perversion: Male-to-Male Sexual Behaviour Outside the West and the Ethnographic Imagination, 1750–1918* (New York: New York University Press, 1995), 11.

44. Petitot, *Quinze Ans sous le Cercle Polaire*, 36.

45. Petitot, *Autour du Grand Lac des Esclaves*, 107.

46. E.g., in his fourth book for the general public, *Autour du Grand Lac des Esclaves*, Petitot compares the mythologies, religions, superstitions, naming practices, and chauvinism of the Dene with those of "civilised" people.

47. Ann Laura Stoler, *Race and the Education of Desire: Foucault's History of Sexuality and the Colonial Order of Things* (Duke University Press: Durham, 1995), 10.

48. Petitot, *Les Grands Esquimaux*, 4–5.

49. Ibid., 53.

50. Ibid., 43.

51. Ibid., 137.

52. Ibid., 283.

53. Ibid., 290.

54. Stoler, *Race and the Education of Desire*, 141.

55. Petitot, *Les Grands Esquimaux*: 24.

56. Petitot, *Autour du Grand Lac des Esclaves*, 103.

57. Petitot, *Les Grands Esquimaux*, 181.

58. Stoler, *Race and the Education of Desire*, 150.

59. Petitot, *Les Grands Esquimaux*, 23.

60. Ibid., 280.

61. Petitot, *En Route pour la Mer Glaciale*, 162.

62. Petitot, *Quinze Ans sous le Cercle Polaire*, 210.

63. Ibid., 232.

64. Petitot, *Quinze Ans sous le Cercle Polaire*, 126, translated it as 'soiled cloth' (*'linge souillé'*) and spelled it "Dzan-you" in *Exploration*.

65. Petitot, *Quinze Ans sous le Cercle Polaire*, 125.

66. Petitot, *Les Grands Esquimaux*, 105.

67. Petitot, *Quinze Ans sous le Cercle Polaire*, 128–130.

68. Séguin to Faraud, Fort Good Hope, September 13, 1866 (RCAY); Petitot to Fabre, Fort Good Hope, September 12, 1866 (AOR).

69. Petitot wrote that Dzan-yu was 20 years old in the spring 1869. See *Exploration*, 113–114.

70. Hyacinthe was the name Petitot used in his correspondence about him.

71. Jeanette Roberts Shumaker, "Abjection and degeneration in Thomas Hardy's 'Barbar of the House of Grebe,'" *College Literature*, Spring 1999: 4.

72. Ali Behdad, *Belated Travelers: Orientalism at the Age of Colonial Dissolution* (Duke University Press: Durham, 1994), 79, cited in Stoler, *Race and the Education of Desire*, 174.

73. Ann Laura Stoler, *Carnal Knowledge and Imperial Power: Race and the Intimate in Colonial Rule* (University of California Press: Berkeley, 2002), 12.

74. Petitot to Faraud, Great Bear Lake, March 17, 1866 (AOR / RCAY); Séguin to Faraud, May 25, 1866 (RCAY).

75. Michel Foucault, *Histoire de la sexualité, Tome 1: La volonté de savoir* (Gallimard: Paris, 1976), 123.

76. Stoler, "Tense and Tender Ties," 35.

77. Petitot to Faraud, Great Bear Lake, May 30, 1868 (AOR / RCAY).

78. Archival letters strongly suggest that Petitot had other liaisons with young Dene men in their mid-teens. Robert Choquette rightly denounced an intimate relationship Petitot had with a "young Indian boy" named Baptiste but did not give his age. See Robert Choquette, *The Oblates Assault on Canada's Northwest* (University of Ottawa: Ottawa, 1995), 61–62. Baptiste was actually a Métis about 12 years old (see Murielle Nagy, "Le désir de l'Autre chez le missionnaire Émile Petitot," in *Sexe et culture dans la rencontre entre Amérindiens et Européens*, ed. Frédéric Laugrand and Gilles Havard (Septentrion: Québec, forthcoming).

79. Choquette, *The Oblates Assault on Canada's Northwest*, 63–66.

80. E.g., Séguin to Faraud, Fort Good Hope, September 16, 1869 (RCAY); Clut to Fabre, Montreal, April 29, 1870 (AOR / AD); Séguin to Taché, Fort Good Hope, May 25, 1871 (AASB).

81. Stoler, *Along the Archival Grain*, 21.

82. Séguin to Faraud, Fort Good Hope, September 16, 1869 (RCAY); Faraud to Fabre, Fort Providence, November 27, 1869 (AOR / AD).

83. Stoler, *Along the Archival Grain*, 252.

84. Petitot, *Exploration*, 319.

85. Choquette, *The Oblates Assault on Canada's Northwest*, 65–66.

86. Petitot asked Bishop Grandin to resign from the Oblates in the spring 1881 (Faraud to Clut, Lake La Biche, November 18, 1882).

87. Choquette, *The Oblates Assault on Canada's Northwest*, 66; Nagy, "Le désir de l'Autre."

88. Ibid.

89. Petitot, *Quinze Ans sous le Cercle Polaire*, xi.

90. Petitot, *En route pour la Mer Glaciale; Exploration*.

91. Pratt, *Imperial Eyes*, 153.

92. Stoler, *Along the Archival Grain*, 40.

IV

IMAGINATION AND COMMODIFICATION

7

———

Marketing Indigenous Bodies in the Fiction of Leslie Marmon Silko, Louise Erdrich, and Sherman Alexie

JOANNA ZIARKOWSKA

"White people want to be Indians. You all have things we don't have. You live at peace with the earth. You are so wise,"[1] says Betty, a white groupie of an Indian rock band, Coyote Springs, in Sherman Alexie's 1995 novel *Reservation Blues*. Betty's words are symptomatic of an intense interest in and enduring fascination with Indian cultures which, as many Native and non-Native critics alike have demonstrated, became an important building block of American identity. Analyzing the phenomenon of appropriating Indian cultural elements, Philip J. Deloria, in his seminal 1998 work *Playing Indian*, observes that what lurks behind the seemingly deep fascination with everything Indian is in fact the manifestation of a desire for a certain interpretation of Indianness rather than Indians themselves.[2] What is sought with unprecedented enthusiasm is not Indians as they are but instead, what they might have been and perhaps never were, a fleeting image, a fantasy of the Noble Savage living in harmony with nature. Yet, as Deloria observes, despite this gross misinterpretation of what constitutes Indianness, Indians have always played a pivotal role in the construction of American identity. "For whites of all classes," Deloria writes, "the quests for personal substance and identity involved forays into racial Otherness,"[3] often performed in an indispensable Indian costume.

This search for authentic American identity anchored in interactions with the Other often took the form of amateurish semi-ethnographic projects. "Object hobbyists" indulged in collecting native artifacts, an activity which endowed its practitioners with the messianic sense that they were preserving a dying culture, and at the same time eliminated the threat of any closer contact

101

with the actual natives. "People hobbyists," on the other hand, preferred direct interactions with the Other and hence created Indian-inspired communes or visited powwows at weekends. These brief and superficial encounters provided a highly satisfying sense of expertise in indigenous culture paired with opportunities to participate in its rituals. While both "object" and "people hobbyists" were interested in different aspects of the Indian Other, they both urgently needed Indianness to construct and authenticate their American identities.[4] A similar conclusion is reached by Shari M. Huhndorf, who in her work *Going Native*, contends that "adopting some vision of Native life in a more permanent way is necessary to regenerate and to maintain European-American racial and national identities."[5] Assuming native costumes and identities, or going native, became a reaction to the abrupt changes of the late nineteenth century, the most significant of which were the rise of industrial capitalism and the completion of the military conquest of Native America. At the same time, the dynamism of rapidly modernizing American society and the ensuing crisis of identity created a sense of nostalgia for origins, best illustrated in the fascination with the primitive. As Huhndorf explains, "idealizing and emulating the primitive, modernity's other, comprised in part a form of escapism from the tumultuous modern world."[6] Hence, both, Deloria and Huhndorf link the insistence on the native component of American identity with the moments of historical crisis which gave rise to collective doubts as to the ways in which social progress was justified and written into national narratives.

While both Deloria and Huhndorf emphasize different aspects of masquerade as a way of establishing collective American identity, this essay explores a corporeal dimension of the phenomenon. Just like native costumes and artifacts, indigenous bodies undergo a similar process of appropriation which transforms them into objects of cultural consumption. Locked in images which refuse to acknowledge contemporary contexts, the body becomes a medium for perpetuating and popularizing a national fantasy of the primitive Other. Through a strategy of reducing Indianness to a set of stereotypical and unrealistic poses, the mainstream culture transforms the indigenous body into a commodity whose circulation on the cultural market is an economically and ideologically lucrative enterprise. The most literal expression of such objectification is the way bodies, in a strategically orchestrated process of dehumanization, are turned into body parts, which are later successfully resold on black markets. This gruesome separation of the body from its human qualities is achieved through a consistent clinicization of vocabulary deployed to refer to the body, which eventually leads to the creation of emotional, mental, and ethical distance facilitating such objectification. Thus, the objectification and commodification of bodies demonstrated in Native American literary texts, apart from symbolically emphasizing the appropriation of Indianness as a form of cultural conquest, culminates in the dehumanization of bodies which become useable and marketable body parts harvested to meet the world's growing demand for human organs.

Louise Erdrich's 1984 novel *Love Medicine* provides a classic example of the appropriation of the indigenous body and its employment in the fantasy of Native Americans as a dying race. Nector Kashpaw, a tribal chairman, recalls his brief career in Hollywood as a film extra. Tall and handsome, he is immediately hired for "the biggest Indian part."[7] However, what playing Indian actually involves is reenacting the fantasy of vanishing Indians since the moment Nector appears in the frame of the shot, he is instructed to "clutch [his] chest" and "fall off that horse." As Nector comments on his film adventure, "That was it. Death was the extent of Indian acting in the movie theater."[8] Disappointed with the film industry, Nector sets out for Kansas, where he receives another unprecedented offer: a rich woman painter asks him to pose for a portrait. What Nector does not realize until the very last moment is that the woman intends to paint him naked: "She put me on a block of wood and then said to me, 'Disrobe.' No one had ever told me to take off my clothes just like that. So I pretended not to understand her. 'What robe?' I asked. 'Disrobe,' she repeated. I stood there naked and confused. Pitiful! I thought. Then she started to demonstrate by clawing at her buttons."[9] When the painting is complete, Nector is astounded by its title and the imagery: *Plunge of the Brave* presents Nector "jumping off a cliff, naked of course, down into a rocky river. Certain death. Remember Custer's saying, 'The only good Indian is a dead Indian?' Well from my dealings with whites I would add to that quote: 'The only interesting Indian is dead, or dying by falling backwards off a horse."[10]

Nector's experience with the white world traces the limits and defines the nature of the Indian presence in American culture. First, Indians are not allowed to control their life narrative: in both film and in painting, the frame and the composition are carefully controlled by white artists, whether filmmakers or painters, who cater to the tastes and preferences of the privileged mainstream. Only when Indians concede and perform the assigned role of vanishing race are they allowed to become part of cultural productions. Second, in the process of appropriation, the indigenous body undergoes an interesting transformation. It is no longer associated with subjectivity and agency but rather becomes a passive and entirely controlled object in a staged presentation of ideologically saturated content. Not only is the body immobilized in artificial poses, fossilized in the historical moment when it amply fulfills a national fantasy about ingenuity, but also, as an object of desire, it is sexualized and eroticized, and thus controlled in yet another context. It is the lean, muscular and attractive indigenous body which becomes the subject of *Plunge of the Brave,* neither the historical reasons for the perilous dive, nor an attempt to accept blame for the annihilation of the brave.

As in Edward Curtis's famous photograph, "Vanishing Race," the Indian body is forever immortalized in the brief moment of passing, disappearing, vanishing. The tragedy of the moment and its historical and ethical implications are glossed over with the aesthetization of the presented subject. The body shown at the moment of perishing becomes objectified and subjected to an indifferent gaze. Separated from the historical and political context of the military and cultural conquest of Native America, the indigenous body acts as an empty signifier, having no counterparts in the real world. Therefore, when Nector looks at the picture, he does not recognize himself in the painted human figure. The effect is the separation of the model from the image whose creation the model enabled. Looking at the image, Nector refuses to identify with the presented vision of Indianness and openly states that he "would fool the pitiful rich woman that painted him and survive the raging water. [He] would hold [his] breath and let the current pull [him] toward the surface, around jagged rocks."[11] His version of the fall, which turns out to be less perilous than expected and clearly subversive in its insistence on survival and thus disappointing for the cultural mainstream, will never become part of the filmed, narrated, or painted story for the simplest possible reason: it does not sell well. As Nector informs us, westerns are popular with mainstream audiences, while *Plunge of the Brave* hangs in the Bismark state capitol. The real Indians, however, those who do not fall off horses and who survive deadly plunges, those who confuse white audiences with their insistence on survival, are not marketable.

That well marketed Indianness can be a lucrative business is amply demonstrated in Sherman Alexie's *Reservation Blues* (1995). The lucrative dimension of Indianness is clearly shown in the invitation that Coyote Springs, an Indian rock band consisting of Junior, Thomas and Victor, Spokane Indians, and Chess and Checkers, Flathead Indians, receive from Cavalry Records, a New York-based music company. The company's agents are attracted by the band's potential to fulfill the fantasy of exotic warriors with warpaint and headdresses. In a fax sent to their home office, it is not the quality of the music that is praised but the alluring qualities of the musicians' ethnicity:

> Checkers is quite striking, beautiful . . . while Chess is pretty. Both would attract men, I think. Sort of that exotic animalistic woman thing. . . . Junior Polatkin is only average on drums but is a very good-looking man. Very ethnically handsome. . . . Overall, the band looks and sounds Indian. They all have dark skin. Chess, Checkers, and Junior all have long hair. Thomas has a big nose, and Victor has many scars.[12]

Obviously, the marketing strategy is based on turning Coyote Springs into what the dominant culture imagines Indians to be and casting the band in the role of the exotic Other, markedly different and yet essential for the construction of the American self. Interestingly, it is physical features that are emphasized as carrying distinctively Indian qualities and being markers of Indian authenticity. The possibility of the band's commercial success rests on the members' appearance which comfortably reinforces rather than questions common stereotypes about Indians: Victor's scars evoke the image of an Indian warrior, Junior, being "ethnically handsome," fulfills aesthetic criteria of Indianness, whereas Chess and Checkers' "animalistic beauty" relegates them to the position of sexually attractive Indian squaws. Indeed, it is Chess and Checkers' ability to attract (white) men that is seen as their biggest asset. The logic of objectification and sexualization of Coyote Springs replicates the logic of cultural conquest, which indiscriminately appropriates a misconstrued understanding of Indianness. In *Reservation Blues*, Cavalry Records and its agents, by no coincidence named Phil Sheridan and George Armstrong, is yet another, contemporary example of how indigenous cultures and bodies are appropriated and commodified to nourish the white man's fantasy.[13]

Interestingly, it is such wide applicability and malleability that becomes the main characteristic of constructed and misinterpreted Indianness. By emphasizing romantic and appealing elements of Indian cultures, such as the connection with Mother Earth or traditional wisdom, Betty and Veronica, Coyote Springs' white, blonde groupies and wannabes, who "[wear] too much Indian jewelry,"[14] represent a highly selective and superficial approach to Native American culture. Like Deloria's "people hobbyists," keen on direct contact with the cultural Other, but not on a daily basis, Betty and Veronica treat Indians as weekend entertainment. Their brief forays into the Spokane Reservation are manifestations of cultural tourism which, as a neo-colonial enterprise, "extracts" the cultural resources needed to resuscitate the connection with the primitive while at the same time consistently refuses to acknowledge less appealing aspects of reservation life, such as poverty, unemployment and alcoholism. Ironically, the trivialization of Indian culture demonstrated by Betty and Veronica, instead of being dismissed as an example of selective appropriation of the most appealing and superficial cultural elements, is enthusiastically embraced by the cultural mainstream as the only approach to Indianness which in fact generates profits. When Coyote Springs fail to secure a contract with Cavalry Records, it is Betty and Veronica who, by relying on their constructed Indian identity, are turned into a product that sells well on the market. As Sheridan explains: "Cavalry Records has an economic need for a *viable* Indian band. . . . Coyote Springs self-destructed. We were thinking we needed a more *reliable* kind of Indian."[15] What reverberates in Sheridan's statement is a desire for Indianness that can be easily controlled and contained in the frame of the fantasy. The apparent ridiculousness of the idea of the two white women performing as an Indian band is dismissed by Sheridan, who explains that Betty and Veronica, by merely having been to the reservation and played with Coyote Springs, have already acquired the necessary amount of authentic Indianness. Physical deficiencies can also be taken care of: "We dress them up a little. Get them into the tanning booth. Darken them up a bit. Maybe a little plastic surgery on those cheekbones."[16] Thus Indianness, as understood by the dominant culture, becomes one of many possible identities to choose from, easy to imitate and cash in on. More importantly, however, authentic Indian looks as defined by the white culture become synonymous with what constitutes Indian identity. Tribal affiliations or ancestry seem of no importance in a context where Indianness is manufactured to be sold for profit. Even the body is evaluated according to how well it fits into the mold of the exotic Indian and, if it fails to satisfy expectations, subjected to necessary alterations, thanks to advancements in plastic surgery.

The indigenous body is a valuable commodity not only in an economic sense but in an ideological sense as well. Russell Kashpaw, an Indian character in Louise Erdrich's 1986 novel *The Beet Queen*, is an example of the Indian body that is first, unscrupulously abused for ideological purposes, and then silenced and eradicated from the cultural discourse. Having served in World War II and the Korean War, white men's wars which he neither supports nor identifies with, Russell, scarred and traumatized, returns to the reservation. Unlike Momaday's Abel, Russell is unable to find either comfort or consolation in the life of an Indian community. According to Louis Owens, by participating in white men's wars, Russell is symbolically embracing the role of an Indian warrior which illustrates the extent of the Indian presence in the discourse of the cultural mainstream.[17] It is only by reenacting the stereotype of a brave Indian warrior that Russell is allowed to participate in the white world and attract its attention. As a military volunteer, he is valuable to America in helping to run its imperialist project. Participation in the white world, however, has its price, as Russell returns from each war more and more scarred, traumatized, and withdrawn. It is the deep and disfiguring scars which he bears on his body that narrate the story of pain and abuse: "The wounds had been so deep that he was ridged like a gullied field. His chest had been plowed like a tractor gone haywire."[18] The agricultural similes that Mary employs to describe Russell's disfigured body bring to mind images of land that has been over-farmed and made barren. Similarly, Russell's body has been subjected to extreme conditions, abused and left uncared for. After returning from the Korean War, Russell gradually loses control over his life, and the stroke that he suffers is the ultimate culmination of the process of becoming disinterested in his life and the people around him, aggravated by the frustration of being trapped in a useless, dysfunctional body.

Immobilized due to partial paralysis, Russell's body is far from useless in the glorious mission of popularizing American ideology. Being "the most-decorated hero" of Argus, Russell is forced to take part in a Beet Festival, an annual town event, as a living albeit motionless symbol of American heroism and endurance:

> The orderly hoisted Russell out of his wheelchair, rolled him onto the bed, and stripped him of his thin cotton pajamas. . . . [Fleur] unpacked Russell's uniform from an old cracked valise. . . . The orderly dressed Russell in it, moving carefully under Fleur's eye. He strained to lift Russell back into his chair. Fleur took Russell's medals from a leather case and pinned the whole bright pattern over his heart. Then she put his rifle, in a long bag of olive drab, across his lap. Russell waited for his hat to be set on at an angle, the way it was in his portrait-studio pictures. . . . The orderly strained to lift Russell onto the float, then strapped him upright between raised wooden bunkers. A field of graves stretched down before him, each covered with plastic grass and red poppies. A plain white cross was planted at his feet.[19]

An event designed to commemorate Russell's participation in American military campaigns is turned into a ridiculous performance in which the celebrated veteran, or rather his paralyzed body, is propped up in a wheelchair for the crowds to observe and scrutinize. Trying to escape the role of a manipulated puppet, Russell alienates himself from the cheering crowd since he is unable to identify with his uniform and the intended symbolism of his posture: he "tried to hold his head high, to keep the fierce gaze smoking, but his chin dropped. His eyes closed, and suddenly the noise and people seemed far off."[20] Unlike Alexie's Coyote Springs, Russell is an example of a *reliable* Indian who, due to his immobility and speechlessness, is easily controlled and integrated into the fabricated narrative of Indianness.

Probably the most gruesome example of the objectification of the indigenous body in the colonial and capitalist context is found in Leslie Marmon Silko's 1991 novel *Almanac of the Dead*. This multilayered text spins the story of myriad characters and covers vast geographical as well as temporal dimensions. The motif of the objectified body is established by Trigg, a Euro-American entrepreneur who suffers a spinal cord injury in an automobile accident. The aim of Trigg's business project is to purchase rundown Tucson real estate and transform the buildings into blood-plasma donor centers, an immensely profitable enterprise in a world faced with the problem of donor shortages and difficulties with the smooth and efficient operation of donor procurement systems. However, in his efforts to obtain blood and donors, Trigg emerges as more than just a resourceful entrepreneur: spurred by racism and classism, he recruits his donors among the homeless and unemployed, illegal Mexican immigrants and poor Indians. The entire population of the dispossessed allows Trigg not only to obtain blood donors but to generate organs for transplants. Thus, the indigenous body, having been objectified and fragmented, in this case literally becomes a product which is resold for profit.

The successful outcome of Trigg's business project rests on a conceptual shift that needs to be introduced, namely the dehumanization of individual bodies and their transformation into spare parts. From the sociological perspective, the shift is by no means easily achieved and accepted since, as Margaret Lock explains,

> body tissues, organs, and even fluids are more often than not, regarded as inalienable, except when made use of in specified ritual practices or in carefully circumscribed activity, such as human reproduction. In order for body parts to be made freely available for exchange they must first be conceptualized as thing-like, as detachable from the body without causing irreparable loss or damage to the individual person or generations to follow.[21]

In Trigg's case, the act of severing the body from its human qualities is a two-stage process. First, Trigg denies the homeless their status as human beings and relegates them to the position of human waste. In his journal, he writes:

> Tucson, city of thieves. . . . These alleged human beings, the filth and scum who pass through the plasma donor center, get paid good money for lying with a needle in their arms—an activity they pursue the rest of the day anyway. I could do the world a favor each week and connect a few of the stinking ones up in the back room and drain them. They will not be missed.[22]

"Discards" and "human debris" are examples of the pragmatic vocabulary that Trigg uses to justify eliminating some, in his view, undesirable categories of people. It is no coincidence, as Ann Folwell Stanford points out, that "it is not *all* bodies that are rendered fodder for scientific and medical gain, but, predictably, those that are deemed worthless (and Other) by the dominant society."[23] The second stage in the objectification process is the substitution of body-as-a-whole imagery with precise and clinical vocabulary which turns coherent bodies into body parts. The terminology associated with waste and debris, initially employed to signal the individual's inferiority in the social hierarchy, is replaced by medical jargon aimed at complete dehumanization of the bodies from which organ-products are extracted. In Trigg's language they are referred to as "donor organs and other valuable human tissue."[24] As "frozen human organs, [are] less reliable, [and] sold for a fraction of freshly harvested hearts and kidneys,"[25] human value is assessed on the basis of how quickly the body can be processed and how many marketable products a single organ harvest will generate. Organs are no longer linked with the individuals from whom they were procured but rather "are transformed into decontextualized objects. Their previous social history is erased, and their value is assessed solely in terms of their quality as organs for transplant."[26] The strictly medical terminology, such as "biomaterials," "the industry's 'preferred' term for fetal-brain material, human kidneys, hearts and lungs, corneas for eye transplants, and human skin for burn victims,"[27] marks the semantic and conceptual shift which facilitates the necessary transition from the body's physical and individual integrity to its total fragmentation and dehumanization.

As improbable as Silko's vision may seem, however, the Indian fantasy and its abuse may be employed subversively to serve the purposes of those whose image it so grossly distorts. The indigenous characters inhabiting Alexie's short fiction are illustrative examples of how Native Americans use and abuse the dominant culture's view on Indian people to meet their own personal needs. In these cases, conforming to stereotypes, especially those referring to Indian appearance, is by no means a gesture of accepting the role assigned by dominant culture but a deliberate strategy to assert agency in the white world. In "Search Engine," a short story from the 2003 collection, *Ten Little Indians*, Corliss, a Spokane Washington State University student, is well aware of how romantic white people tend to be about Indians, and constructs her public image based on the exoticism and mystery that seem so appealing to her white fellow students:

> If white folks assumed she was serene and spiritual and wise simply because she was Indian, and thought she was special based on those mistaken assumptions, then Corliss saw no reasons to contradict them. The world is a competitive place, and a poor Indian girl needs all the advantages she can get. So if George W. Bush, a man who possessed no remarkable distinctions other than being the son of a former U.S. president, could also become president, then Corliss figured she could certainly benefit from positive ethnic stereotypes and not feel guilty about it.[28]

Edgar, a Spokane Indian professional, from "Class," a short story from Alexie's 2000 collection *The Toughest Indian in the World*, uses Indian stereotypes as his flirting strategy. Bearing in mind a set of associations that his brown muscular body evokes, Edgar deliberately casts himself in the role of the noble and therefore sensuous Indian, thus fulfilling the fantasy of the return to the primitive: "I'd told any number of white women that I was part Aztec and I'd told a few I was completely Aztec. That gave me some mystery, some ethnic weight, a history of glorious color and mass executions. Strangely enough, there are aphrodisiacal benefits to claiming to be descended from ritual cannibals."[29] Not only does Edgar reverse the effect of the ethnic stereotype but also ridicules and takes advantage of the white obsession with the concept of Indian authenticity: if a Spokane body is not Indian enough, then an Aztec one certainly is. Corliss and Edgar are examples of individuals who know very well what the dominant culture expects them to be and, by playing out the fantasy, they manage to follow their own agendas. Moreover, as a lawyer, Edgar, relies on the power of stereotypes to sway jurors' opinions in the courtroom: again, knowing how to fit his body into the stereotype, he keeps his hair long and braided and thus evokes a myriad of feelings, from a fascination with the ethnic Other to a sense of guilt due to the historical consequences of the "discovery" of America. The interesting dimension of such treatment of Indian stereotypes in Alexie's short fiction is, despite its seemingly oppressive nature, a reversal of the roles in the paradigm: it is Indians who directly benefit from the existence of stereotypes. White people remain completely oblivious to the fact that this time it is they who are being exploited, ironically with the use of the master's tools. As interesting as Alexie's subversion of stereotypes is, it does not escape criticism about its limited scope. Corliss and Edgar may profit from the existence of ethnic and racial stereotypes, but they will never introduce the Indian presence on their own terms and control its content. The paradigm remains the same, except for the fact that this time Indian characters also profit from its existence. However, despite these shortcomings in Alexie's picture of contemporary Indian life, it is worth noticing that his narratives introduce Indian characters who are no longer victims of historical circumstances, unable to control their lives, but rather individuals who adequately cope with the demands of a life dominated by a fantasy which so grossly distorts their cultural and ethnic image.

"Indian is easy to fake. People have been faking it for five hundred years,"[30] says Harlan Atwater, an urban Indian and poet sought by Corliss in Seattle. The use of the pronoun "it" is no coincidence here: in spite of his reference to "faking" Indianness, Harlan is really addressing Indianness that is imagined and contracted by the dominant culture. The act of "faking" Indianness involves the appropriation of not only native bodies but also ethnic costumes, rituals, spirituality, traditional medicine and other cultural practices, which inevitably reduces the appropriated cultures to the status of a commodity. As a result of this process, the merging of political, economic and ideological interests produce cultural imperialism, which, in Laurie Anne Whitt's words, is "a form of oppression exerted by a dominant society upon other cultures . . . which secures and deepens the subordinated status of those cultures." Whitt contends that cultural imperialism "undermines [Native American] . . . integrity and distinctiveness, assimilating them to the dominant culture by seizing and processing vital cultural resources, then remaking them in the image and marketplaces of the dominant culture."[31]

The corporeal aspect of the process of remaking as discussed by Native American writers is an illustration of yet another dimension of this objectification. The process of objectification rests on several ideologically-oriented strategies: immortalizing Native Americans in artificial images of the past which deny the agency and subjectivity of the presented subjects; trivialization of Indian identity, which becomes equivalent with a set of stereotypical physical features and whose thorough knowledge is secured by brief weekend encounters reminiscent of cultural tourism; sexualization of the body, seen as readily available for the gaze (and pleasure) of the cultural mainstream; the deployment of the disabled and silenced indigenous body as an example of the triumph of imperialism; and, finally, the dehumanization of the body and its transformation into saleable body parts. Subjected to the logic of the cultural and economic marketplace, indigenous bodies become a tool of banishing Native American voices from American narratives and thus sustaining the practice of cultural imperialism. There are, however, gestures of resistance. Acknowledging the existence of objectification and the superficial understanding of what constitutes Indianness, Native American characters consciously and methodically comply with stereotypes manufactured by the cultural mainstream in order to follow their own agendas. Thus, the usurpation of Indian identity, so insightfully commented on by Deloria and Huhndorf, is challenged through a subversive dialogue with the fantasy.

Notes

1. Sherman Alexie, *Reservation Blues* (New York: Grove Press, 1995), 168.
2. Philip J. Deloria, *Playing Indian* (New Haven: Yale University Press, 1998), 90.
3. Ibid., 132.
4. Ibid., 135–41.
5. Shari M. Huhndorf, *Going Native: Indians in the American Cultural Imagination* (Ithaca: Cornell University Press, 2001), 8.
6. Ibid., 14.
7. Louise Erdrich, *Love Medicine* (New York: Harper Perennial, 1993), 123.
8. Ibid., 123.
9. Ibid.
10. Ibid., 124.
11. Ibid.
12. Alexie, *Reservation Blues*, 190.

13. James Cox, "Muting White Noise: The Subversion of Popular Culture Narratives of Conquest in Sherman Alexie's Fiction," *SAIL* 9.4 (1997): 62.

14. Alexie, *Reservation Blues,* 41.

15. Ibid., 272. Emphasis added.

16. Ibid., 269.

17. Louis Owens, *Other Destinies: Understanding the American Indian Novel* (Norman: University of Oklahoma Press, 1992), 210.

18. Louise Erdrich, *The Beet Queen* (New York: Harper Perennial, 2006), 74–5.

19. Ibid., 298–9.

20. Ibid., 299.

21. Margaret Lock, "Alienation of Body Parts and the Biopolitics of Immortalized Cell Lines," in *Beyond the Body Proper: Reading the Anthropology of Material Life,* ed. Margaret Locke and Judith Farquhar (Durham: Duke University Press, 2007), 572.

22. Leslie Marmon Silko, *Almanac of the Dead* (New York: Penguin Books, 1992), 386.

23. Ann Folwell Stanford, *Bodies in a Broken World: Women Novelists of Color & the Politics of Medicine* (Chapel Hill: The University of North Carolina Press, 2003), 186. Emphasis in the original.

24. Silko, *Almanac,* 387.

25. Ibid., 404.

26. Margaret Lock, *Twice Dead: Organ Transplants and the Reinvention of Death* (Berkeley: University of California Press, 2002), 49.

27. Silko, *Almanac,* 398.

28. Sherman Alexie, "Search Engine" in *Ten Little Indians* (New York: Grove Press, 2003), 11.

29. Sherman Alexie, "Class" in *The Toughest Indian in the World* (New York: Grove Press, 2000), 40.

30. Alexie, *Ten Little Indians,* 40.

31. Laurie Anne Whitt, "Cultural Imperialism and the Marketing of Native America" in *Native and Academics: Researching and Writing about American Indians,* ed. Devon A. Mihesuah (Lincoln: University of Nebraska, 1998), 141–2.

8

Stories from the Womb—
Esther Belin's *From the Belly of My Beauty*

Ewelina Bańka

You don't have anything
if you don't have stories.
. . .
He rubbed his belly.
I keep them here
[he said]
Here, put your hand on it
See, it is moving.
There is life here for the people.

And in the belly of this story
The rituals and the ceremony
are still growing.[1]

Through the image of a storyteller in whose body, as if in a mother's womb, stories grow, Leslie Marmon Silko points to the fundamental role played by storytelling and language in Native American tradition, which is "to heal, to regenerate, and to create."[2] Using the "reproductive" power of language to shape stories, the storyteller becomes an "agent of cultural continuity"[3] and, thereby, performs a role which is at the heart of Native community, namely, that of a mother—a life giver and nurturer. Perceived as a living memory of the community, the storyteller shapes the historical consciousness of the people becoming a medium transmitting the entire knowledge to successive generations. In doing so, she/he incorporates the life of the community into the body of a mythic story that, according to traditional Native thought, goes back to the origins, "deep into the womb of the earth, back to the beginning

113

[of life], to the first underworld."[4] Emerging from the earth, the unwinding story is believed to reinforce the relation between the land and its inhabitants. As Simon Ortiz claims, understood as a spiritual energy that comes from the natural world,[5] language and story are central in the process of shaping Native identities and express an intimate, kinship relation held between the people and the first mother: the earth.

The arrival of Europeans in the Western Hemisphere resulted in the systematic violation of the bond between the land and Native people which, consequently, led to the fragmentation of many Native communities. As Ortiz states, the loss of lands to the emerging nation-state has been "[t]he greatest and most horrible trauma Indigenous people of the Americas have experienced and endured for hundreds of years since European settlement, colonization, [and] conquest."[6] European and later American settlements have for years been absorbing vast areas of indigenous lands, dispossessing Native communities of their spiritual homelands. The removal of indigenous people from their lands triggered a long and complex process of cultural fragmentation and estrangement since the loss of territories means, in reality, deprivation of landscapes that traditionally informed Native people's faith as well as their identity. As Ortiz stresses, this continuous fragmentation of indigenous communities equals both physical and psychic displacement that affects contemporary Native people living on the reservation homelands as well as in the urban centers. In the light of this cultural change Ortiz points to the need to redefine the question of identity-formation and the meaning of "home" in contemporary Native American, often urban, experience.

The process of negotiating home as well as the self is explored in a book of poetry *From the Belly of My Beauty* by Navajo/Diné poet Esther G. Belin. Raised by her Diné parents in Lynwood, California, educated at the University of California at Berkley and the Institute of American Indian Arts in Santa Fe, New Mexico, Belin represents a second generation of off-reservation Native people, "URIs. (Urban-Raised Indians)," "city cousins,"[7] forming a multitribal community constantly growing in number across the United States. As Belin states, her poetry is deployed to plant "the seed of revolution"[8] in response to the dominant forces of Western culture that keep Native people "[i]n a state of colonial confusion."[9] Composed of poems and an autobiographical essay, the bilingual collection forms a poetic commentary on urban Indian struggles in modern-day America seen through the poet's eyes, where her intimate stories are interwoven with the experience of California's multitribal urban community as well as the stories of the Diné people living on the reservation homeland.

Central to Belin's artistic project is the use of body imagery which is deeply grounded in the traditional Diné system of knowledge. The subsequent analysis will demonstrate how Belin employs the image of the body to graphically present Native people's struggle and resistance to the colonial fragmentation of indigenous lands and people as well as to demonstrate the process of reclaiming home and an indigenous sense of self in contemporary Native experience. The Native body becomes in Belin's poems a site of resistance to racial discrimination, social marginalization, and cultural estrangement. Using the image of the womb, Belin reconnects with the matrilineal culture of her Diné mother and celebrates the strength of indigenous motherhood. This reconnection, perceived by the poet as a form of homecoming, leads to Belin's growing self-awareness as a contemporary Diné mother and poet. Consequently, her poetic story from the womb becomes an act of life-giving which "create[s] what lives: survival/of colored peoples in this country/called the United States."[10] Furthermore, identifying the Native body with the earth, Belin reestablishes her connection with the body of Dinétah—Navajo homeland. Eventually, portrayed as an extension of indigenous land, the Native body in Belin's poetry manifests the beauty, strength and, most of all, the survival of Diné people.

Moving Homeward: Learning the Diné Ways

The first part of the collection comprises poems that manifest the narrator's feeling of cultural estrangement in the urban milieu and her attempts to incorporate herself into the communal body of Diné community. The narrator's struggle in the process of self-identification as a contemporary urban Diné woman, and her coming to terms with urban America as home, mirror the struggles of a larger multitribal Indian community, growing in the urban areas across the country. In the first lines of the poem opening the collection Belin describes the place from which her voice and her story emerge:

> The beckon for liberation itches on my
> back. I scratch desert-dried arroyos littered
> with fading Budweiser cans and roadside
> crosses dangling plastic flowers. I stumble
> from the grotesque reflection recycled. I
> stumble with my commod-filled body
> running a Wheaties race. I stumble at my
> shadow raised by Los Angeles skyscrapers.[11]

The feeling of heaviness, confusion, and insecurity, intensified by the word "stumble," repeated almost like a mantra, dominates in the first part of the poem. Identifying her overburdened "commod-filled body" with the industrial landscape of urban California, the poet blurs the boundary between her own body and the body of the land. Thus the poem foreshadows the story unwinding on the streets of California which is about the Diné poet, the people she represents and, inevitably, about the land.

Gravitating between the streets of L.A. and Dinétah, Belin learns to overcome the feeling of physical dislocation and cultural alienation, incorporating the body of Diné teachings and the spirit of Diné culture with her urban academic upbringing which has "[r]e-rout[ed her] tribal identity with capitalist influences."[12] She realizes that her overwhelming sense of alienation is a result of her exposure to Western institutions which have taught her that "Indian land was far away in another world, across states lines."[13] Following the teachings of her grandparents, she begins to look at the land through the prism of Diné worldview that defies the concept of geopolitical borders imposed externally on indigenous communities. Learning the Diné ways the poet realizes that although living off the reservation, she is physically within the body of her people's homeland as "L.A. has sacred mountains"[14] that constitute part of Dinétah.

Fragmentation of the body of indigenous lands is mirrored in the fragmentation of Native communities and, consequently, of Native identities. The mosaic of Indian voices emerging from Belin's poems gives testimony to joys and hardships of city life, braiding together the stories of home, family, and love with the accounts of urban displacement, racism, alcohol abuse, violence, or fragmented identities. The voices heard in Indian bars tell stories of relocation, "rooted for invasion/imperial in destiny,"[15] that gave birth to the ever-growing urban community of the "dislocated funky brown."[16] As Janet McAdams claims,[17] the parallel between the image of the fragmented land and the fragmented self is demonstrated through Belin's use of numbers, such as in the example of the poem entitled "Ruby in Me # 1:"

> middle child
> smart child
>   ¼ Navajo
>   ¼ Navajo
>   ¼ Navajo
>   ¼ Navajo
>
> four parts equal my whole
> # 311,990
>
> enrolled = proof
> 50
> 80
> 100 if you can stand
> it

veiled
minority status
alcohol
resemblance[18]

Composed of a list of short definitions and numbers inserted in-between, the poem's structure points to the writer's will and yet, inability to define her identity when she is forced to comply with different criteria of "authentic Indianness," constructed as a compilation of various "parts" such as blood quantum, tribal enrolment, or federal recognition.

Writing the poetic story becomes therefore a quest for the indigenous sense of self that, according to Navajo thought, is understood as an inseparable part of a larger, collective body—the Diné nation. A poem "Check One," which in its form resembles a questionnaire for ethnic identity, alludes to racial identity politics, leading to mental confinement of the indigenous mind, as the mechanisms of Native peoples' self-definition have for years been subjugated to and controlled by the outside world:

Check One:

Diné   ☐
Other   ☐[19]

Challenging the reader to tick the appropriate answer, Belin points to the racial discrimination and artificiality inherent in such systems of representation, and demonstrates that one of the factors leading to the fragmentation of people and cultures is the imposed necessity to choose one's identity by a simultaneous physical differentiating/separating oneself from the "other." By inviting the reader to choose between the two options, the poet positions herself as subjected to the external forces of racial representation. Her "vulnerability" alludes to the situation of contemporary Native people who, embedded in the political body of the nation-state, are forced to grapple not only with the geopolitical borders imposed on their homelands but also with "blood borders," created by a legal system that controls identity definitions through externally imposed norms and categories. Therefore the process of homecoming becomes for the poet an act of liberating herself from "a contrived reality boxed into *Indian*/ Identifying the branches of soul wounds/into another contrived reality called/ *American* AKA United States."[20]

The tension present in the collection is balanced with poems devoted to the stories of Indian mothers, which, together, form the main thread in the narrative of Native communities striving "in this/country called the United/States."[21] It is not without reason that these stories constitute such

an important part of the collection since central to Belin's process of reincorporating herself within the body of Diné community and Dinétah is the
reconnection with her Diné mother.

"My Mother Is My Story"[22]

With this declaration, which opens the essay included in the book, Belin
manifests her conscious reintegration with the body of Diné matrilineal and
matrilocal culture of her mother. The "amá/mother story" is at the heart of
the collection, playing a fundamental role in the poet's growing self-under-
standing as a Diné woman and a Diné poet-storyteller. It is simultaneously a
declaration of her reclaiming Navajo language/Diné bizaad—"[t]he passage to
[her] existence"—which, informing and shaping her sense of self, empowers
her to "take the role of little sister./ . . . of Zia[23] woman/ and mother."[24] Thus
the Diné bizaad becomes, to use Trinh Minh-ha's terminology, "the language
of her entrails,"[25] which shapes the story in the poet's womb, and enables
her to fully understand her role as a Native woman and a poet-storyteller.
Reclaiming her tribal language is therefore an act of homecoming, which,
as Inés Hernández-Avila contends, takes place in a process of "relocating of
our languages in the homes of our words, and our homes in the words of
our languages."[26]

What informs Belin's process of reclaiming the Navajo language is also
the awareness of her mother's struggles over the years spent at a boarding
school, away from the reservation homeland, which focused on the "annihilation of savage tendencies characteristic of indigenous peoples" and on teaching them "to silence your native tongue, voice, [and] being."[27] Acknowledging
her mother's daily struggles as a relocated urban Diné living away from the
reservation homeland, Belin honors her by incorporating the "amá story" into
her own poetic revolution:

Re-Entry

My mother is my story.
   She sacrificed for me, allowing me to use the enemy's tongue.
Perhaps to reverse the process. Perhaps to change the process.
Perhaps so that I could survive the process easier than she.

   *To acknowledge that I can manipulate the English language is to tell
my her-story or re-tell shimá.*

   To acknowledge that I can manipulate the English language is to say
my tribal language is scrambled within me. In my blood silently circulating.
In my back pocket squashed incomprehensible. The color of my skin. The

rhythm, ba-bum, the ticking, ba-bum, the map in my heart, ba-bum, leading me home.[28]

Circulating in Belin's blood, the Navajo language becomes for the poet a story and a map of the land which, through the figure of the Diné mother, holds the people into one body, the body of the Diné nation. The process of reclaiming the language, thus learning to read the map, starts from analytical deciphering a system of Diné sounds, letters and words that Belin identifies when listening to the elders. However, with the poet's growing self-awareness, the Diné bizaad changes into a living, bodily presence which, through the routes of her veins, leads her back home, allowing her to describe and identify herself with "a known, remembered, imagined [and] longed-for terrain"[29] of Dinétah.

Recreating a map in/of her body, Belin restores balance in her life by reclaiming the relation with her homeland. As Janice Gould argues, this map-making, identified in the work of many Native American women writers and poets, is "a reflection of the need to know and love our Mother, to repair our bond with her, and through her, with all our Indian family, all our relations."[30] Formed by her tribal language, Belin's poetry becomes thereby an "organic, nurturing writing,"[31] and an expression of her growing consciousness as a Diné woman-mother-poet. Since the primary role of a woman in Diné society is that of a nurturer, the fundamental responsibility of a mother focuses on providing sustenance for her children. Therefore, as Maureen Trudelle Schwarz observes, "[o]n the basis of this link, the concept of mother extends to all factors in the world that contribute to the sustenance and development of human life . . . virtually anything that contributes to sustenance and development is a mother in the Navajo world."[32] Thereby, Belin's poetic writing, which is simultaneously the "amá story," becomes the mother's "skin"—"a body capable of receiving as well as giving: nurturing and protecting."[33]

Guiding Belin homewards and affirming her place within the Diné community, the Diné bizaad becomes a tool that "clarifies [the poet's] resistance"[34] to the body of dominant culture which for years has displaced, marginalized and silenced indigenous people. It becomes a means of her poetic revolution carried out in a form of "wordarrows"[35] that unwind the story of survival of Native people in the "brick-wall sanctuar[ies]"[36] of urban America. Eventually, her story forms a body interwoven with a larger narrative by the Third World women writer-storytellers. Belin's stories about "the survival [of colored people] under colonial rule"[37] become tools in the process of the liberation of her people from the exploitation of their minds, bodies and lands. Pointing to the role of the Third World women's narratives, Trinh Minh-ha argues that a woman writer who works at un-learning the dominant language of "civilized" missionaries also has to learn how to un-write and write anew. And she often does so by re-establishing the contact with her foremothers, so that living tradition can never congeal into fixed forms, so that life keeps on nurturing life, so that what is understood as the Past continues to provide the link for the Present and the Future.[38]

The stories emerging through the voice of Belin's poems remain a continuation of her Diné mother's story, which, essentially, is a story of life that emerges through the body of the mother-storyteller. Empowered by the language rooted in Dinétah, as well as by the reconnection and identification with amá, Belin's revolutionary story is created:

II
This is from the center of
my skull where all is cosmic
and glacier washed from
reading journals of cultural
research called *Paris Review*
and *onthebus*. From the
melodic muse of my belly I
create what lives: survival
of colored peoples in this
country called the United
States. The cosmos, meteor-
showered with oppression,
cramps my back and balls
my fists, smashing the glass
ceiling against my nose,
bony and intricate are those
stories.
From Ruby she wails[39]

Emerging from the center of her skull as well as from her womb, Belin's writing resembles a child/birth process.[40] Bearing the whole cosmos inside her, the Diné poet-mother creates and "gives birth" to the world and people molded with words inside her body, becoming thereby a universal mother and, consequently, a universal storyteller, for, as Minh-ha states, "every woman is the woman of all women."[41] Ruby, the main character of the poems included in the middle section of the collection, can be interpreted as the poet's twin or her relative, her child, her friend, or her double. Thus, as Janet McAdams observes, representing various women (be it a mother, a lover, a teenage girl, or a new born baby), Ruby becomes "an Indian Everywoman."[42] In "The Laugh of the Medusa" Hélène Cixous claims that in the act of writing, a woman puts her "self" into the text, simultaneously reclaiming her place in the history and in the world.[43] Imagining Ruby's life, the poet-mother writes herself, as well as Native women and their people, into the text that documents various fragments of indigenous history ignored in the dominant narrative of American nation. The collection becomes therefore a celebration of indigenous womanhood and an affirmation of the survival of Native people.

Woman—Mother—Earth

The roots of "amá/mother story" in Belin's poetry go beyond the Zia clan of her mother's people and reconnect the poet with yet another body—the earth. Since for the Diné the word "amá" refers not only to a mother by birth, but to the earth as well,[44] the narrator's "organic" story becomes inevitably connected with the Navajo story of creation of the first people in the womb of the earth. Perceived by the Navajo as "small in size, a floating island in mist or water,"[45] the First World, called "the World of Darkness and Dampness,"[46] resembles the womb, the place of origins of the first people, who without definite forms yet (like a fetus in a womb) were awaiting to be shaped into humans.

The concept of an organic kinship between the people and the earth constitutes a direct challenge to the notion of "blood borders" that Belin learns to resist. The theme of reclaiming the bond with the earth is seen in the poem entitled "Bringing Hannah Home." The portrayal of the poet digging a hole in the ground on the Navajo Reservation, in order to bury the newborn Hannah's placenta in it, highlights the Navajo's belief in the physical and spiritual kinship bond between the people and the land. The act of bringing the baby girl home refers not only to the process of returning to the reservation homeland. Digging a hole in the ground to bury the placenta, the poet acknowledges the nurturing power of the earth which is believed to be "life" that will "nourish" and "embody" Hannah.[47] Returned to the ground, "to become one with Mother Earth again,"[48] the placenta epitomizes the blood relation between the earth and her people. It is also a symbolic aligning between the mothers for, as Paula Gunn Allen indicates, the body of the woman (represented by the placenta) is the body of the earth herself; it is a planet with life flourishing in as well as on it.[49]

While burying the placenta in the ground the poet recalls yet another ceremony, notably the burial of her late father which she also perceives as an act of "homecoming."[50] Therefore the juxtaposition of the two ceremonies becomes the poet's manifestation of her belief in the organic relation held between the people and the earth, within whose body human life emerges and ends. As Allen argues, reclaiming one's relation with the earth means "Walking in balance . . . knowing that living and dying are twin beings, gifts of our mother, the Earth."[51] Thus the poem becomes another "amá story," reconnecting "womb to womb, body to body,"[52] in which the woman-mother-earth nourishes and guides her people on their journey of life. Taking the form of "continuous multiple pregnancies"[53] that create stories from the womb, Belin's story becomes rooted in the body of the earth-home and, woven together with the story of mythic creation,

> draws its corporeal fluidity from images of water [the water present in the First World]—a water from the source, a deep, subterranean water that trickles from the womb, a meandering river, a flow of life, of words running over or slowly dripping down the pages. This keeping-alive and life-giving water exists simultaneously as the writer's ink, the mother's milk, the woman's blood and menstruation.[54]

Through her organic, life-giving writing, the Diné poet aligns herself with the narrative of human creation, carrying in her womb a story of the universal journey of humanity. This poetic act of homecoming reconnects Belin symbolically with the body of yet another amá—Changing Woman—considered by the Navajo as the mother of all people.[55]

Asdzáá Naadlee'hí/Changing Woman

For the Navajo, Changing Woman is a holy person who "came to the world to save it"[56] and who created the first Diné people. She is identified as "the source and sustenance of all life on the earth's surface, controlling particularly fertility and fecundity."[57] According to Diné teachings, Changing Woman created the first Nihookáá Dine'é (People of this World—the Fourth World) by rubbing her skin wastes from her breasts, her back, sides, and from under her arms, mixing it with corn as well as with white shell, abalone, turquoise and jet—the four fundamental components of Diné cosmology.[58] She used the material to create the first Diné and gave life to them with her own breath and her voice by talking or singing to them. Since all that is alive in the world is constructed of the same substance, the human body, created by Changing Woman, is an inseparable organic part of the body of the universe. Having guided the first people to Dinétah, Changing Woman taught them about the natural order in the world and about who they are as people.[59] Then, together with other Holy People, the mother of all people became part of the land, giving the world to the humans.[60] Therefore, according to the traditional Diné thought, the Navajo are bodily connected with Mother Earth, being "the flesh and the seed of the Holy People."[61]

In the process of her growing self-awareness as a Diné mother-poet, Belin creates a story that ensures the survival of "human existence on this planet."[62] In its "reproductive" power the poet's story mirrors the process of creation of the first Diné by Changing Woman. The language of the poems is Belin's "linguistic skin," used to create and nurture the story of the emergence of the multitribal Indian nation scattered across the country. Stressing the organic character of language, Gloria Anzaldúa argues that the spirit of the words that move in and emerge from the writer's body is "as concrete as flesh and as palpable."[63] The poetic words of Belin's story, like Changing Woman's song, are used to (re)create and sustain the people and their culture. Thus, through her work, the poet takes the role of a mythic creatrix—a "shaper of existence in [her] tribe and on the earth."[64]

Reconnecting with and continuing the "amá story," the urban Diné mother-poet-creatrix continues her poetic narrative, sewing together and thus creating the body of a community which comprises the "mixedblood, crossblood, fullblood, urban, rez, relocated, terminated, non-status, tribally enrolled, federally recognized, non-federally recognized, alcoholic, battered, uranium-infested."[65] Creating a vision of Native survival, the poet reconnects "the dislocated funky brown"[66] into a body of a multitribal family and creates home for them in the language of her poetry. This act becomes a process of the poet's own homecoming and reclaiming a sense of self as a contemporary Diné. As Hernández-Avila claims, "Each time a Native woman seeks, unearths, brings forth, and defends her people's history, political identity, and cultural teachings, she finds increasingly discrete parts of herself."[67]

Returning Home

Reclaiming her connection with the body of Dinétah and the matrilineal culture of her mother, Belin takes on the role of a storyteller whose poetic revolution is to nurture and affirm "human existence on this planet."[68] As Simon Ortiz claims, the language of poetry is "an energy that forms us and also at the same time is the essence of how we come into being."[69] Belin's nurturing story, centered on the figure of amá, becomes therefore a map that guides contemporary Native people on their journey homewards toward cultural self-awareness that, according to Native thought, they started in the womb of the earth as one people. As the journey continues, turning into "[an ironic] immigration to the place some call the United States,"[70] Belin's "creation story" affirms and celebrates indigenous continuance. Eventually, in the process of returning home and reclaiming her Diné self, Belin finds her own place within the body/circle of Native American women writers. From the heart and womb of this body emerges the nurturing "amá story" which contains "maps to deep truth(s), maps that help us to know the lay of the land of our bodies, of our points of origin/emergence, of our hearts and spirits, of the universe, of our minds, of the planet we call home, we call Earth."[71]

Notes

1. Leslie Marmon Silko, *Ceremony* (New York: Penguin Books, 1977), 2.

2. Joy Harjo, "Introduction." *Reinventing the Enemy's Language. Contemporary Native Women's Writing of North America,* eds. Joy Harjo and Gloria Bird (New York: W. W. Norton and Company, 1997), 21.

3. Mary Chapman, " 'The Belly of This Story': Storytelling and Symbolic Birth in Native American Fiction." *Studies in American Indian Literatures* 7, no. 2 (Summer 1995): 3.

4. Maureen Trudelle Schwarz, *Molded In the Image of Changing Woman. Navajo Views On the Human Body and Personhood* (Tucson: The University of Arizona Press, 1997), 18.

5. Simon J. Ortiz, "Indigenous Language Consciousness: Being Place, and Sovereignty." *Sovereign Bones. New Native American Writing,* ed. Eric Gansworth (New York: Nation Books, 2007), 135.

6. Ibid., 140.

7. Esther G. Belin, *From the Belly of My Beauty* (Tucson: The University of Arizona Press, 1999), 74.

8. Ibid., 71.

9. Ibid., 72.

10. Ibid., 23.

11. Ibid., 1.

12. Ibid., 71.

13. Ibid., 9.

14. Ibid., 9.

15. Ibid., 11.

16. Ibid., 3.

17. Janet McAdams, "Review: Histories of the World. *Indian Cartography* by Deborah Miranda, *From the Belly of My Beauty* by Esther G. Belin, *Blue Marrow* by Louise Bernice Halfe." *The Women's Review of Books* 17, no. 10/11 (July 2000): 44–46.

18. Belin, *From the Belly,* 39.

19. Ibid., 12.

20. Ibid., 1.

21. Ibid., 23.

22. Ibid., 67.

23. Zia (Tł'ógí) is the name of the family clan of Belin's mother.

24. Ibid., 18.

25. Trinh Minh-ha, *Woman, Native, Other: Writing Postcoloniality and Feminism* (Bloomington: Indiana University Press, 1989), 37.

26. Inés Hernández-Avila, "Relocations Upon Relocations: Home, Language and Native American Women's Writing." In *Reading Native American Women: Critical/Creative Representations,* ed. Inés Hernández-Avila (Lanham: Altamira Press, 2005), 173.

27. Belin, *From the Belly,* 68

28. Ibid., 67.

29. Janice Gould, "Poems as Maps in American Indian Women's Writing." In *Speak to me Words. Essays on Contemporary American Indian Poetry,* eds. Dean Rader and Janice Gould (Tucson: The University of Arizona Press, 2003), 22.

30. Ibid., 25.

31. Minh-ha, *Woman, Native, Other,* 38.

32. Schwarz, *Molded In the Image of Changing Woman,* 27.

33. Minh-ha, *Woman, Native, Other,* 27.

34. Cherríe Moraga's poem "It's the Poverty" quoted in Gloria Anzaldúa, "Speaking In Tongues: A Letter to 3rd World Women Writers." In *This Bridge Called My Back. Writings by Radical Women of Color,* eds. Cherríe Moraga and Gloria Anzaldúa (New York: Kitchen Table: Women of Color Press, 1983), 166.

35. A term coined by Gerald Vizenor in *Wordarrows: Indians and Whites in the New Fur Trade* (Minneapolis: University of Minnesota Press, 1978).

36. Belin, *From the Belly,* 61.

37. Ibid., 71.

38. Minh-ha, *Woman, Native, Other,* 148–149.

39. Belin, *From the Belly,* 23.

40. Minh-ha, *Woman, Native, Other,* 37.

41. Ibid., 135.

42. McAdams, "Review: Histories of the World," 45.

43. Hélène Cixous, "The Laugh of Medusa," 347.

44. In Navajo, the word amá refers to "a mother by birth, the earth, the sheep herd, the corn field and the mountain soil bundle." See: Gary Witherspoon, *Language and Art in the Navajo Universe* (Ann Arbor: University Press, 1977), 91.

45. Aileen O'Bryan, *Navajo Indian Myths* (New York: Dover Publications, Inc., 1993), 2.

46. Ibid., 4.

47. Belin, *From the Belly,* 13–14.

48. Schwarz, *Molded In the Image of Changing Woman,* 138.

49. Paula Gunn Allen, *Off the Reservation: Reflections on Boundary-Busting, Border-Crossing, Loose Canons* (Boston: Beacon Press, 1998), 118–119.

50. Belin, *From the Belly,* 13–14.

51. Paula Gunn Allen, *Off the Reservation. Reflections on Boundary-Busting, Border-Crossing, Loose Canons* (Boston: Beacon Press, 1998), 118–119.

52. Minh-ha, *Woman, Native, Other,* 136.

53. Anzaldúa, *Borderlands: La Frontera. The New Mestiza* (San Francisco: Aunt Lute Books, 1999), 96.

54. Minh-ha, *Woman, Native, Other,* 38.

55. Schwarz, *Molded In the Image of Changing Woman,* 24.

56. Laura Tohe, "Changing Women." In *Sister Nations. Native American Women Writers on Community,* eds. Heid Erdrich and Laura Tohe (St. Paul: Minnesota Historical Society, 2002), 3.

57. Schwarz, *Molded In the Image of Changing Woman,* 24.

58. Ibid., 31.

59. Ibid.

60. For different versions of the creation of the first Nihookáá Dine'é see: Aileen O'Bryan, *Navajo Indian Myths* (New York: Dover Publications, Inc., 1993), Washington Matthews, *Navajo Legends* (Salt Lake City: University of Utah Press, 1994), Leland Wyman, *Blessingway* (Tuscon: University of Arizona Press, 1970).

61. Schwarz, *Molded In the Image of Changing Woman,* 63.

62. Belin, *From the* Belly, v.

63. Gloria Anzaldúa, *Borderlands,* 93.

64. Paula Gunn Allen, *The Sacred Hoop. Recovering the Feminine in American Indian Traditions* (Boston: Beacon Press, 1986), 31.

65. Belin, *From the Belly,* 77.

66. Ibid., 3.

67. Hernández-Avila, "Introduction: 'It Is What Keeps Us Sisters:' Indigenous Women and the Power of Story." *Frontiers: A Journal of Women Studies* 23, no. 2 (2002): 2.

68. Belin, *From the Belly,* v.

69. David Dunaway, "Simon Ortiz. The *Writing the Southwest* Interview. July 14, 1988."In *Simon J. Ortiz. A Poetic Legacy of Continuance,* eds. Susan Berry Brill de Ramírez and Evelina Zuni Lucero (Albuquerque: University of New Mexico Press, 2009), 99.

70. Belin, *From the Belly,* 3.

71. Inés Hernández-Avila, "Introduction," x.

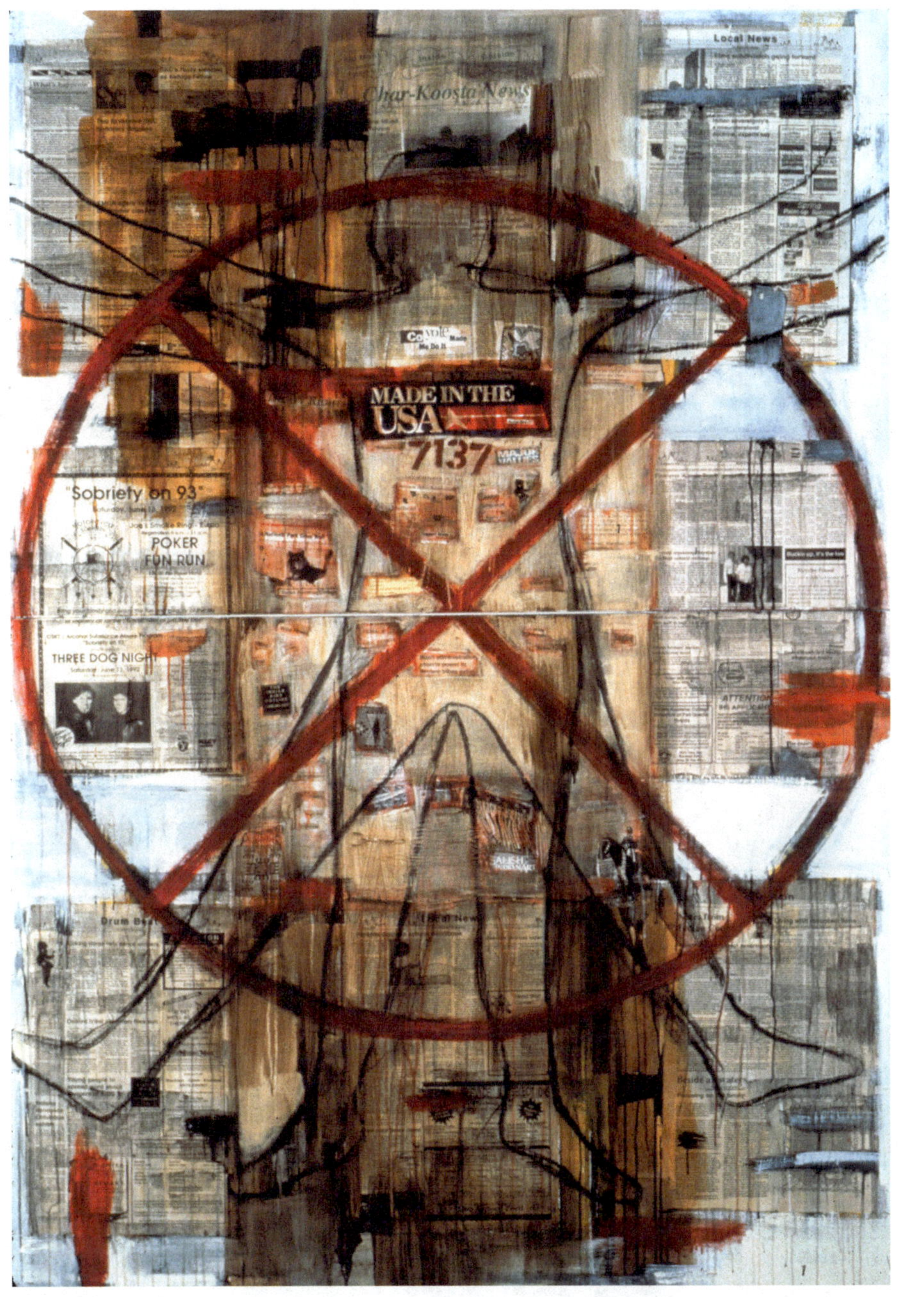

Figure 1. Jaune Quick-to-See Smith, *The Red Mean: Self-Portrait*, 1992, acrylic, newspaper collage, shellac, and mixed media on two canvas panels, 90 × 60 inches. *Credit:* Smith College Museum of Art. Part gift from Janet Wright Ketcham, class of 1953, and part purchase with the Janet Wright Ketcham, class of 1953, Fund, ID Number: SC 1993:10a, b.

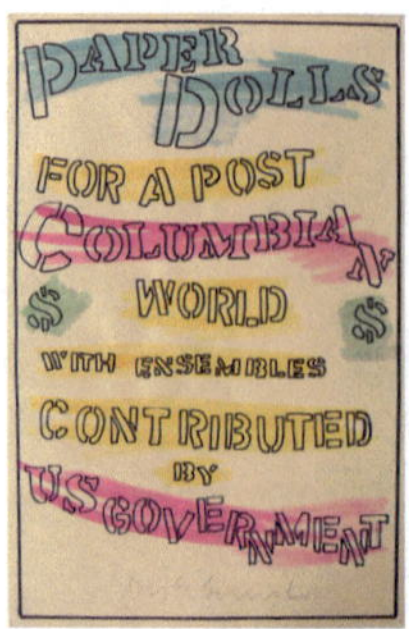

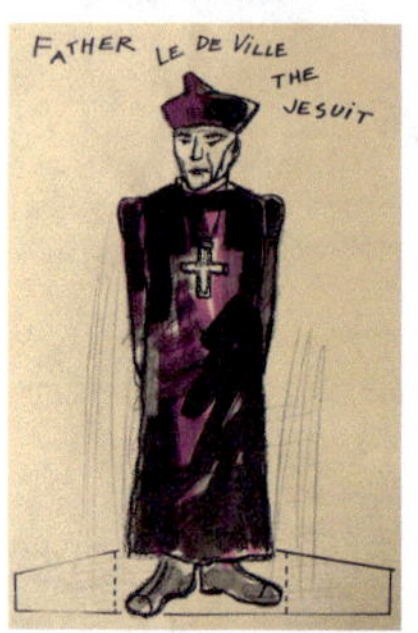

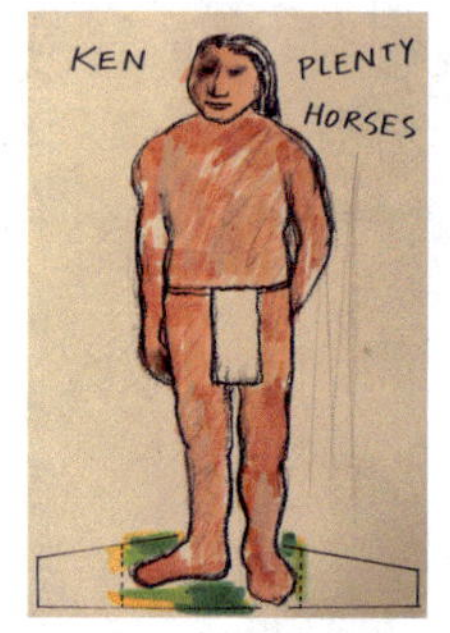

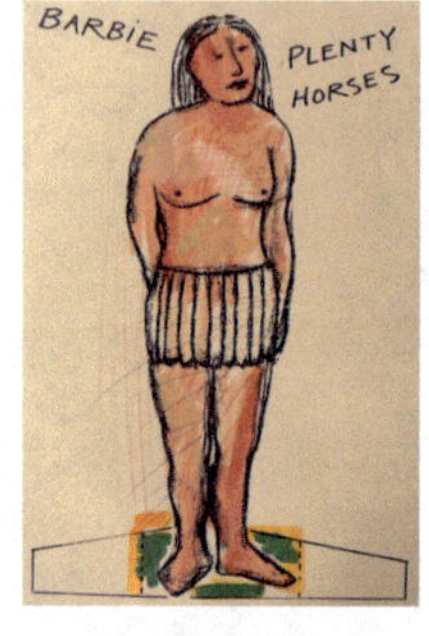

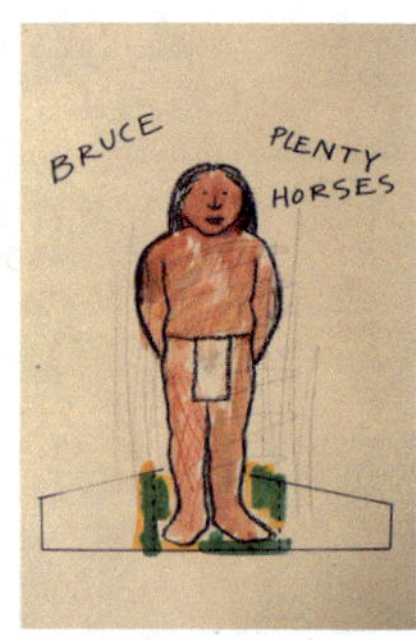

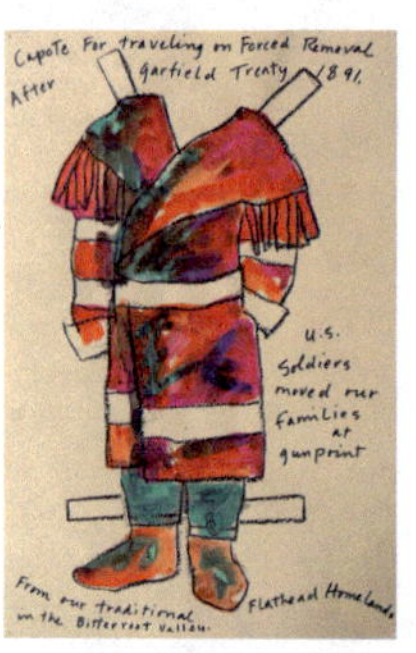

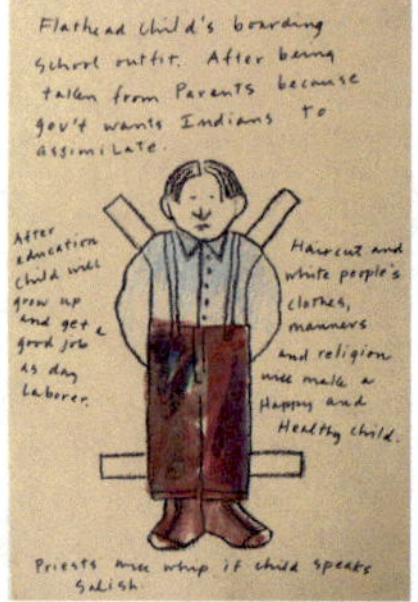

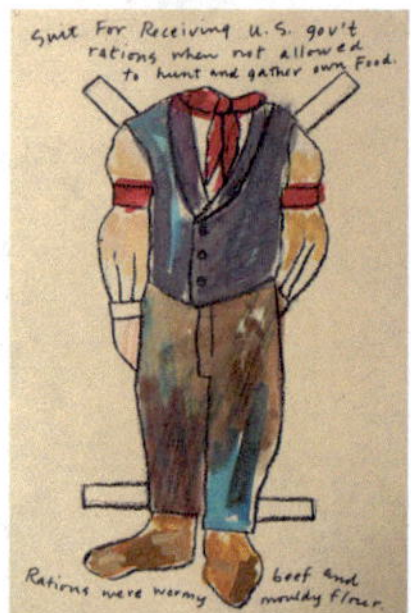

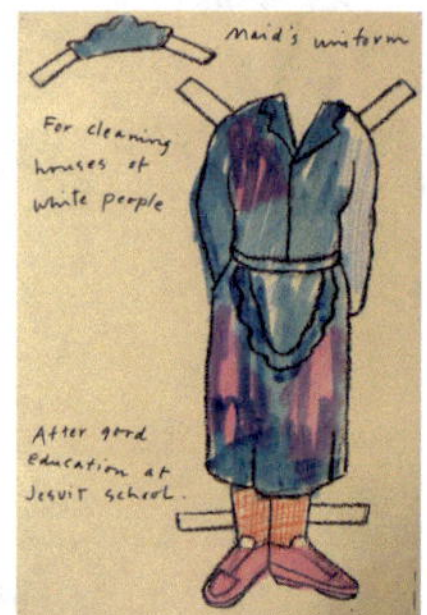

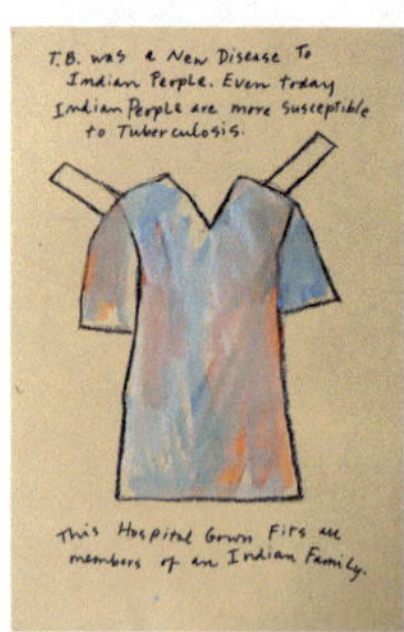

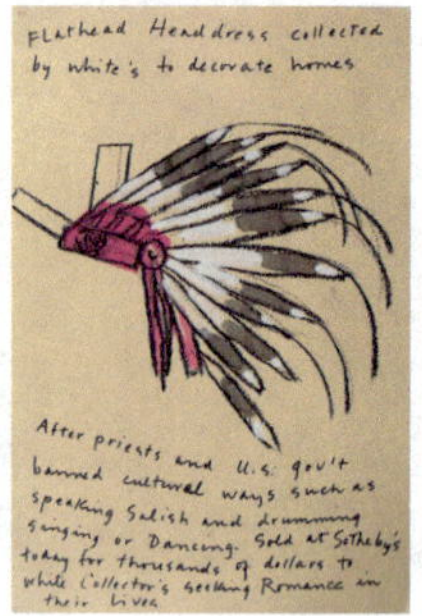

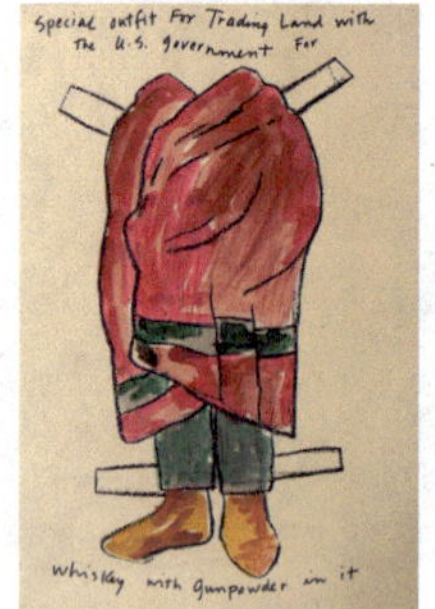

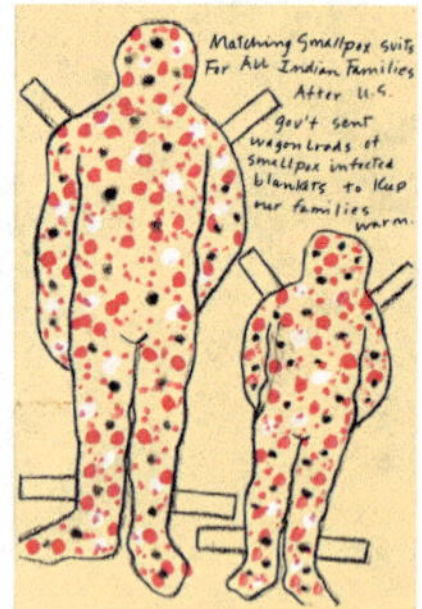

Figure 2. Jaune Quick-to-See Smith, *Paper Dolls for a Post-Columbian World with Ensembles Contributed by the U.S. Government*, 1991–1992, thirteen watercolor, pen, and pencil on photocopy paper, each measuring 17 × 11 inches. *Credit:* New Mexico Museum of Art. Gift of Lucy R. Lippard, 1999 (1999.15.301.1).

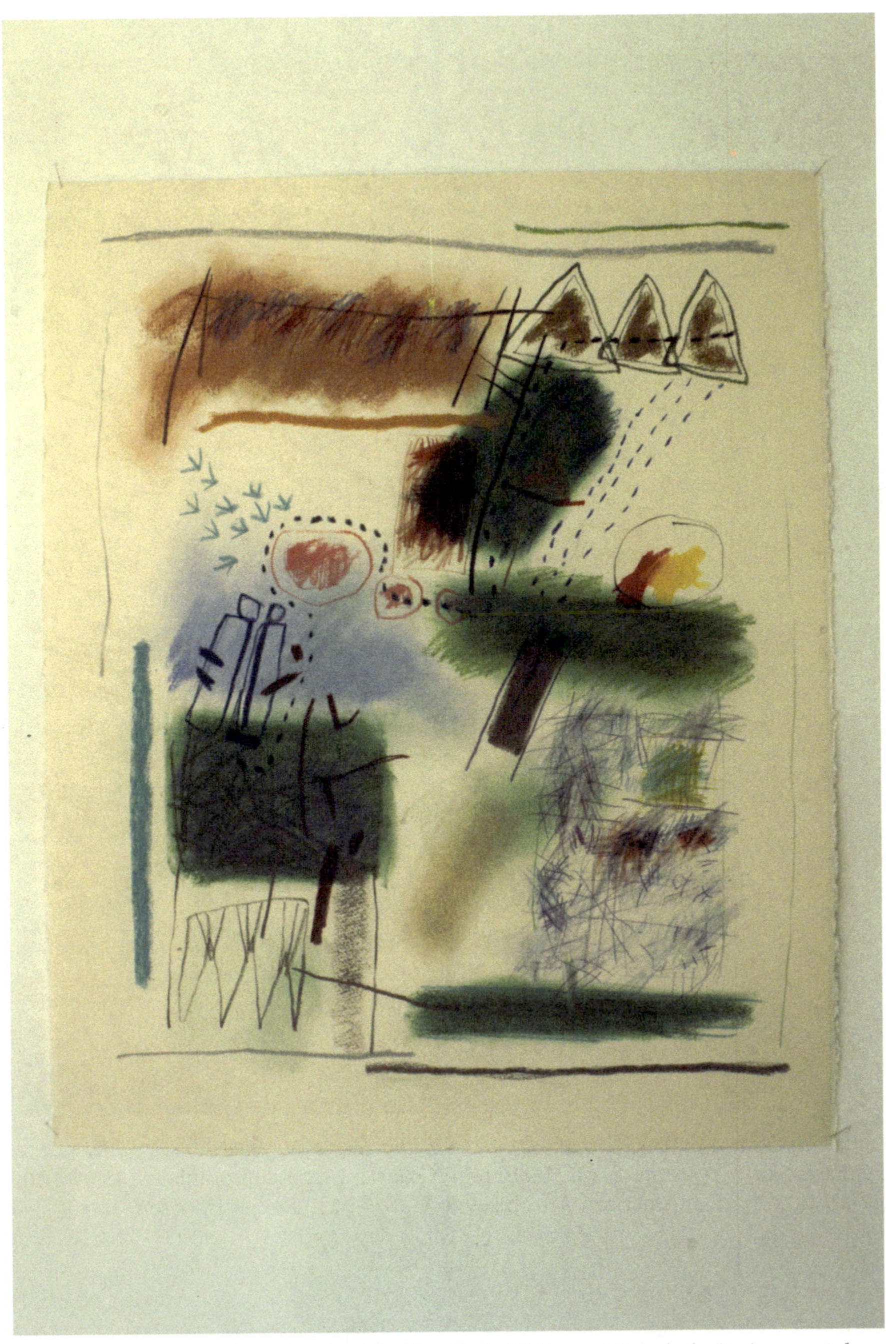

Figure 3. Jaune Quick-to-See Smith, Untitled, *Wallowa Waterhole* Series, pastel on paper, 41¼ × 35⅛ inches. 1978. Courtesy of Jaune Quick-to-See Smith.

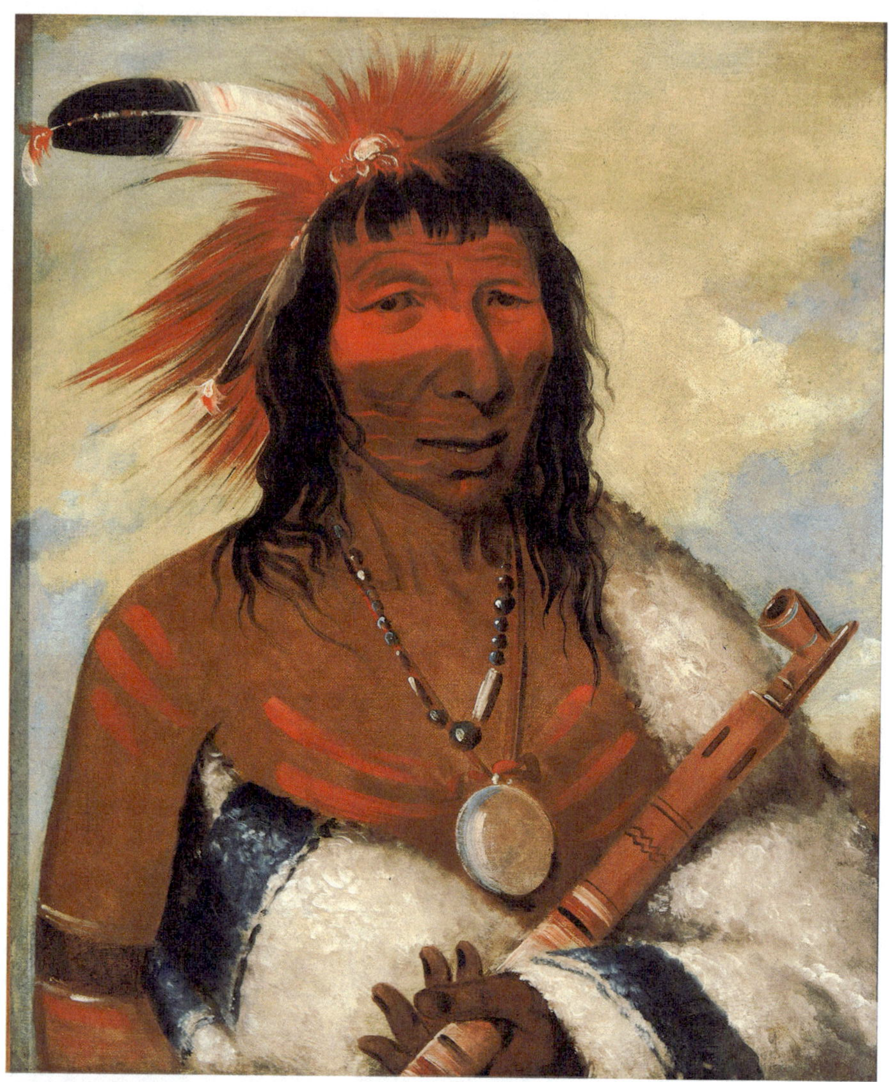

Figure 4. George Catlin, *Wa-nah-de-tunk-ah, Big Eagle (or Black Dog), Chief of the O-hah-kas-ka-toh-y-an-te Band*, 1835, oil on canvas (Accession number: 1985.66.70). *Credit:* Smithsonian American Art Museum. Gift of Mrs. Joseph Harrison, Jr.

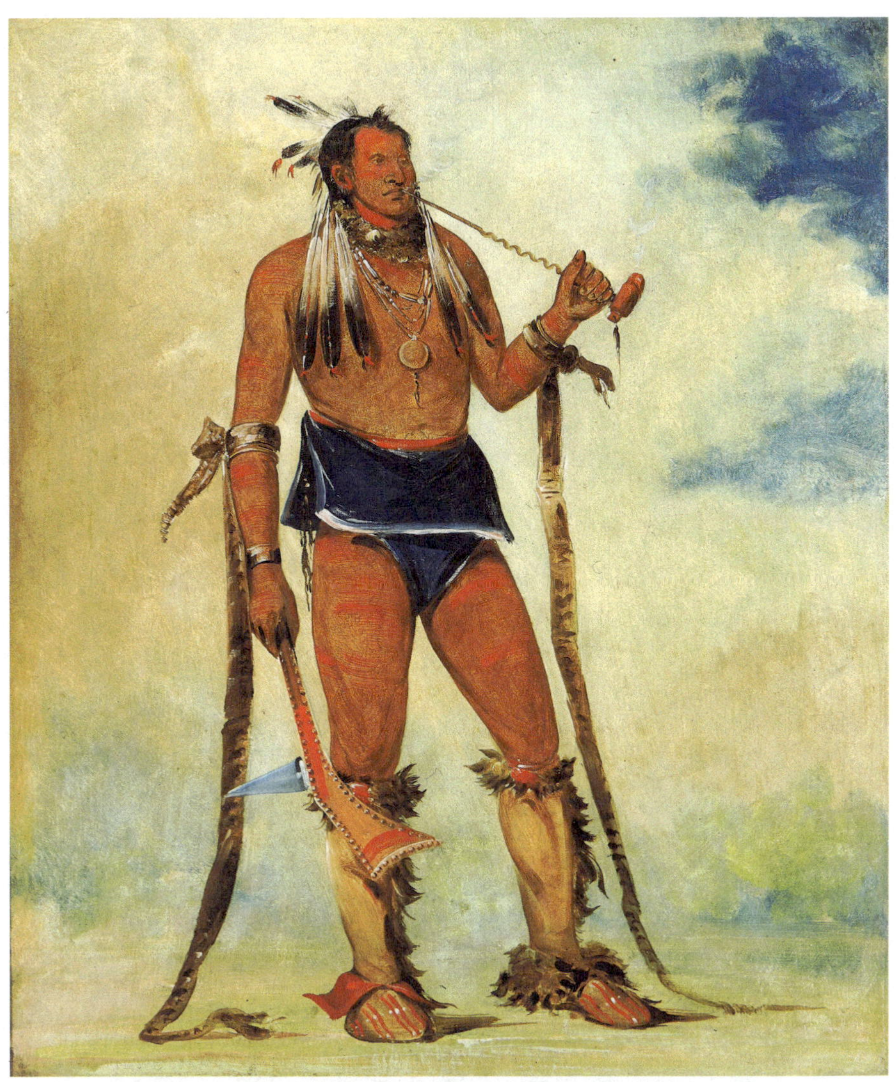

Figure 5. George Catlin, *Wah-chee-háhs-ka, Man Who Puts All Out of Doors*, 1835 (Accession number: 1985.66.207). *Credit:* Smithsonian American Art Museum. Gift of Mrs. Joseph Harrison, Jr.

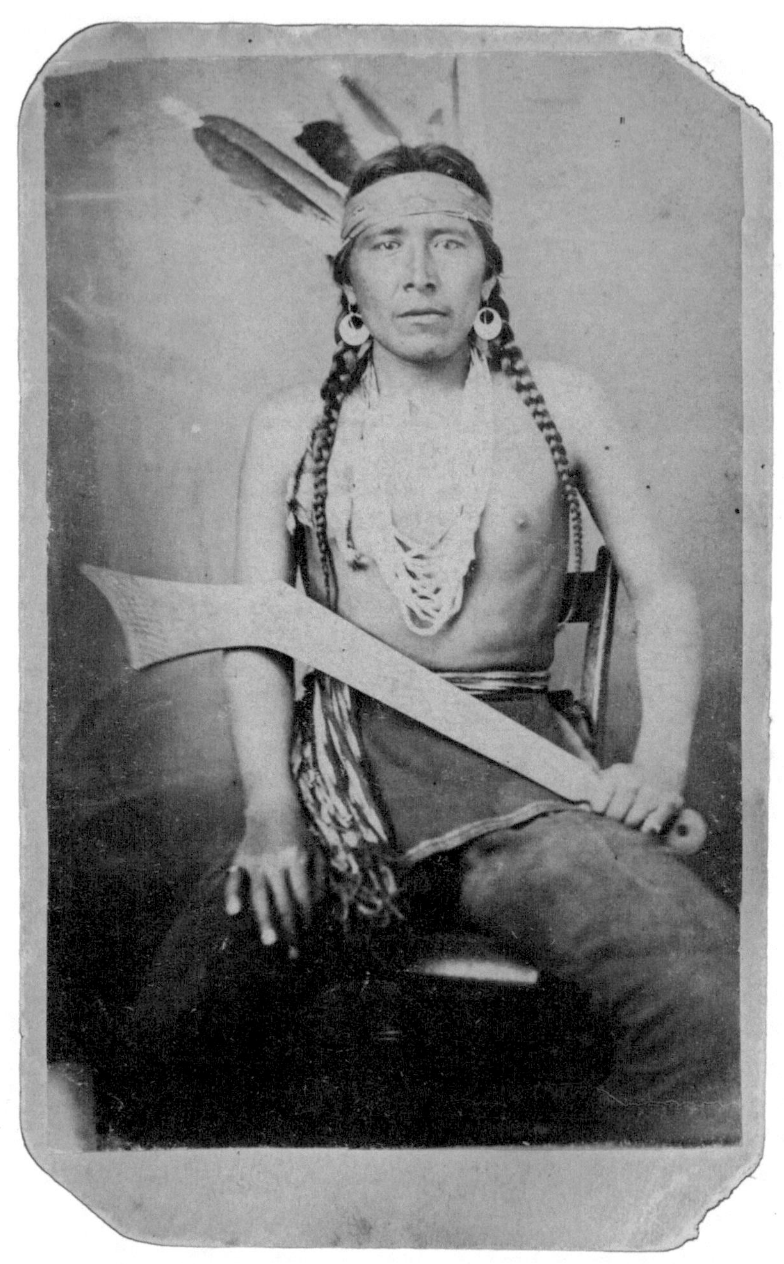

Figure 6. *Wamditanka, Jerome Big Eagle,* c. 1864 photograph by W. W. Hathaway. *Credit:* Courtesy of the Minnesota Historical Society.

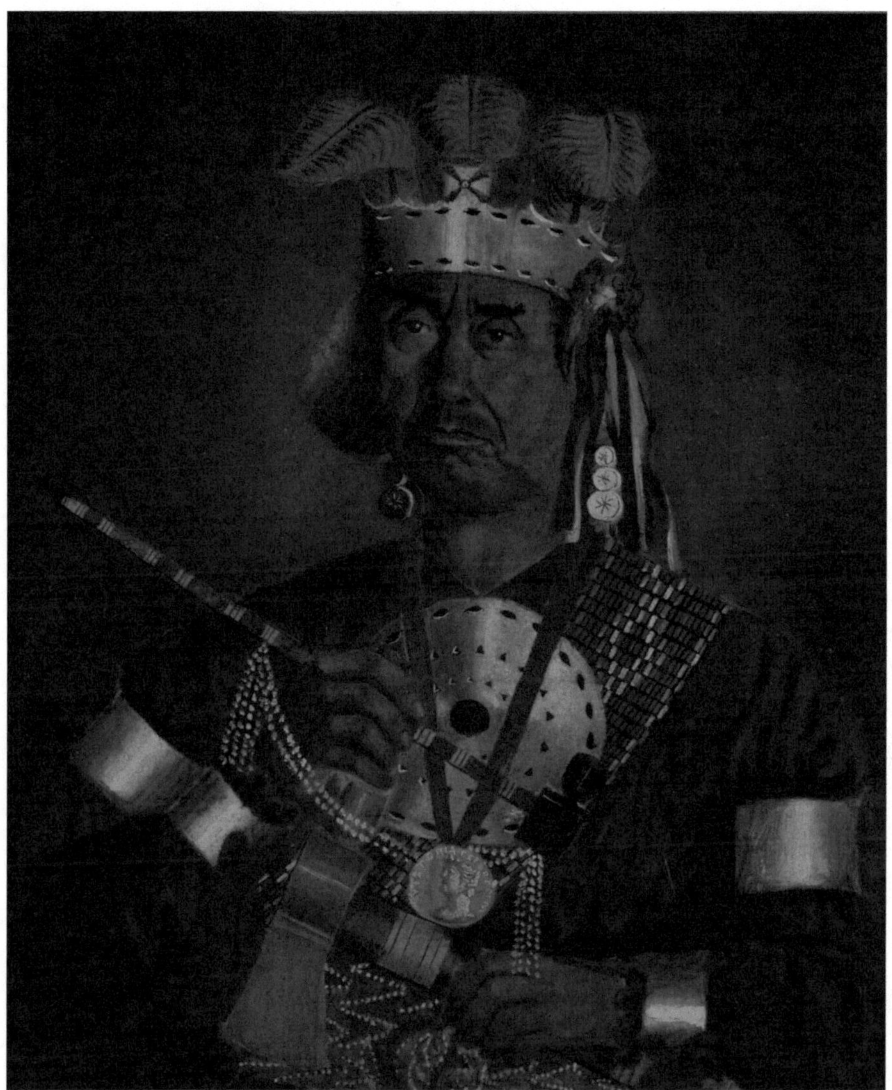

Figure 7. Zacharie Vincent (Huron of Lorette) Self-Portrait, mid-nineteenth century, oil on canvas. *Credit:* Musée de la civilisation, collection du Séminaire de Québec. Restrauration effectuee par le centre de conservation du Québec, photograph Pierre Soulard (accession number: 1991.102).

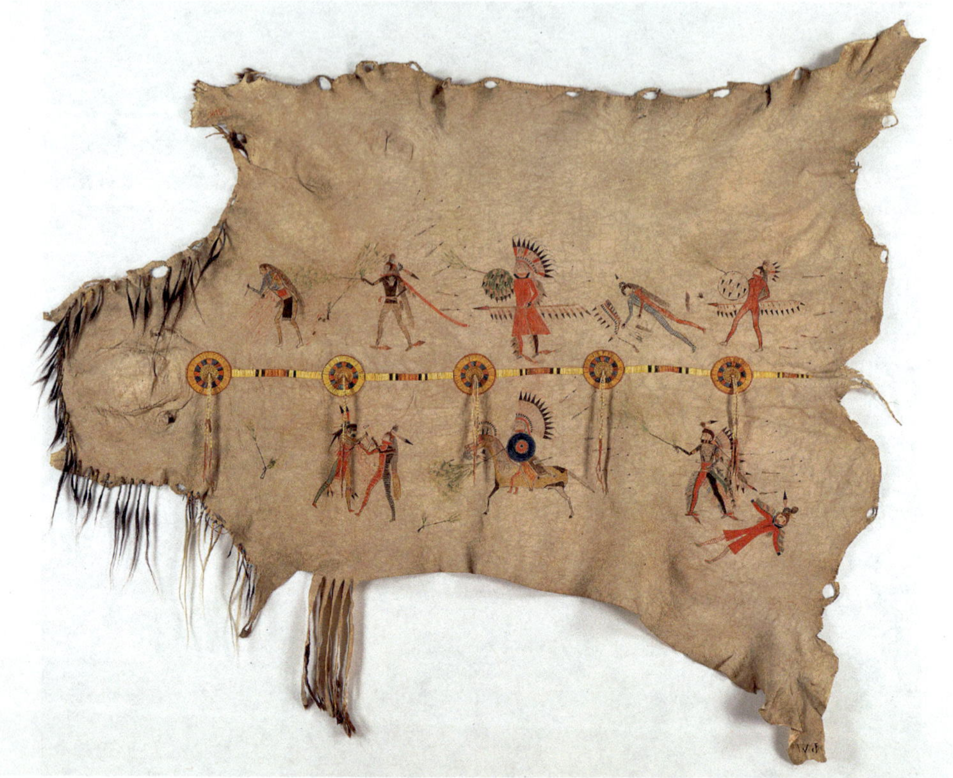

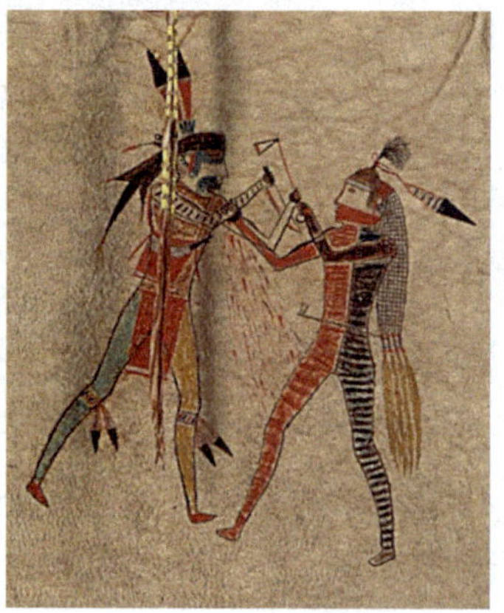

Figure 8. Mato-tope (Mandan), painted bison robe of the 'second' chief of the Mandan, Mato Topé ("Four Bears"), c. 1830s. *Credit:* © Linden-Museum, Stuttgart. Photo: A. Dreyer. Enlarged detail, of bottom left-hand figures on robe, shows Mato-tope's pipe painted horizontally across his waist.

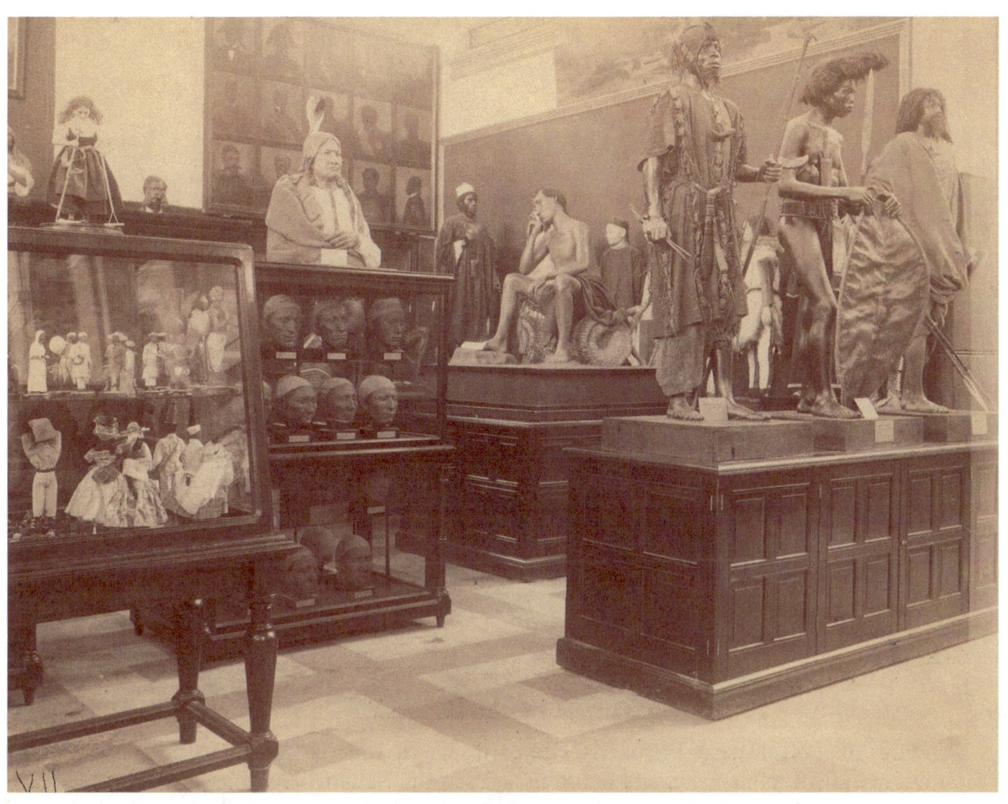

Figure 9. Fort Marion Prisoners' Heads, displayed with mannequins and Sitting Bull bust, in the National Museum c. 1890, NAA, Smithsonian Institution, Photo Lot 4 09707900. *Credit:* National Anthropological Archives, Smithsonian Institution.

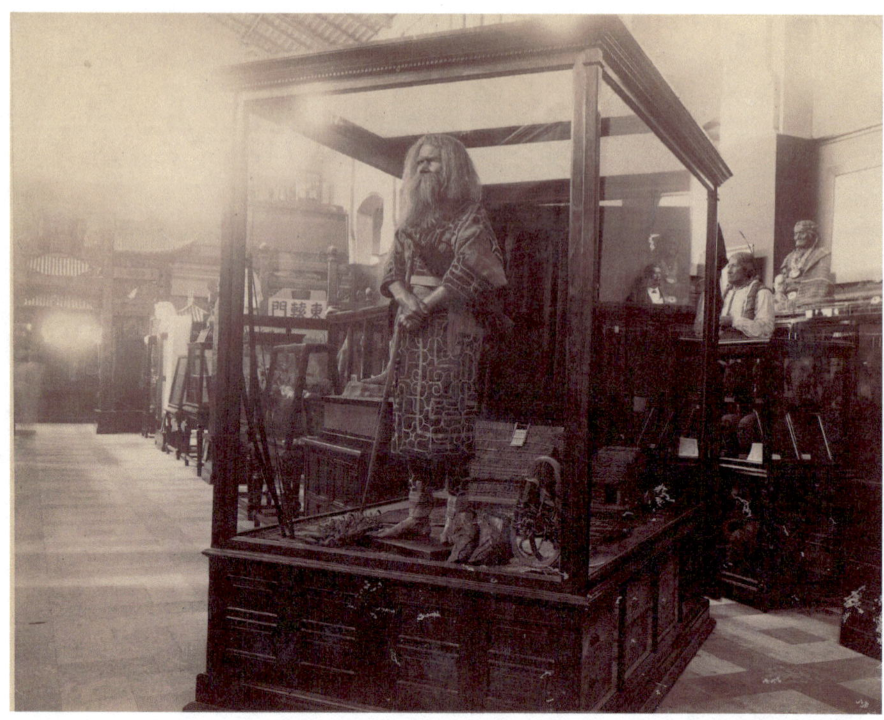

Figure 10. Fort Marion Prisoners' Heads, displayed with costumed mannequin of Ainu man with implements and house models, in the National Museum c. 1890, NAA, Smithsonian Institution, Photo Lot 4 09708400. *Credit:* National Anthropological Archives, Smithsonian Institution.

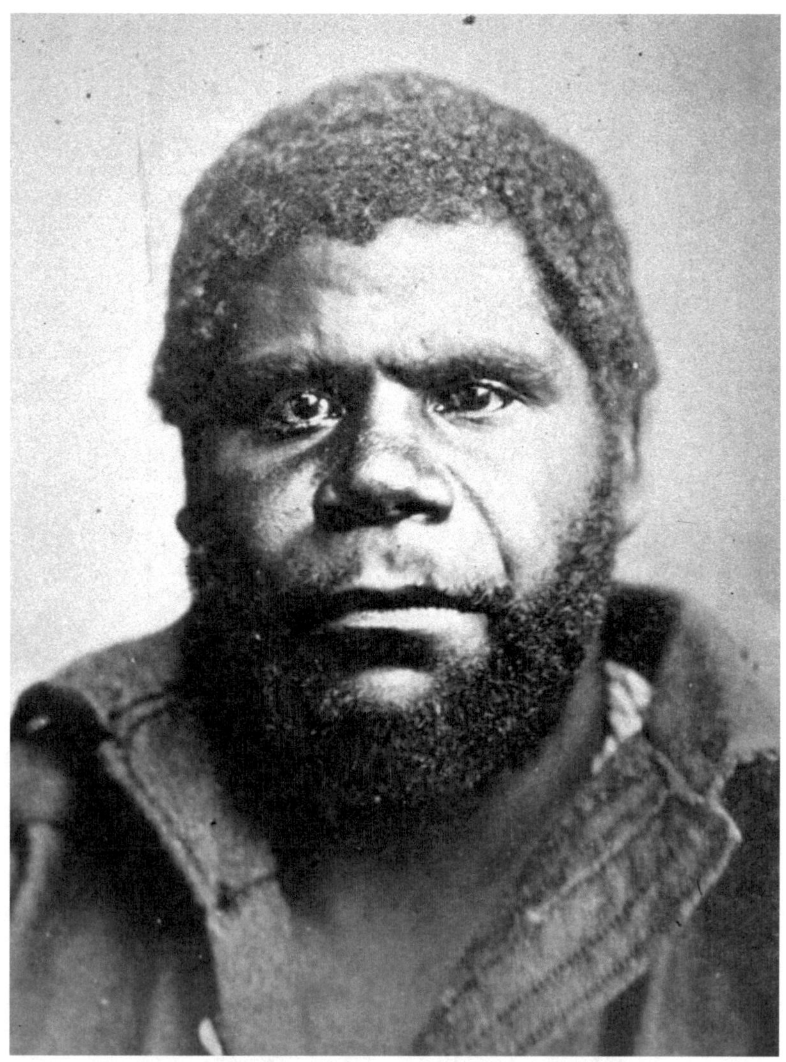

Figure 11. Charles Alfred Woolley, *Carte de visite of William Lanné (King Billy)*, 1866. *Credit:* State Library of Tasmania.

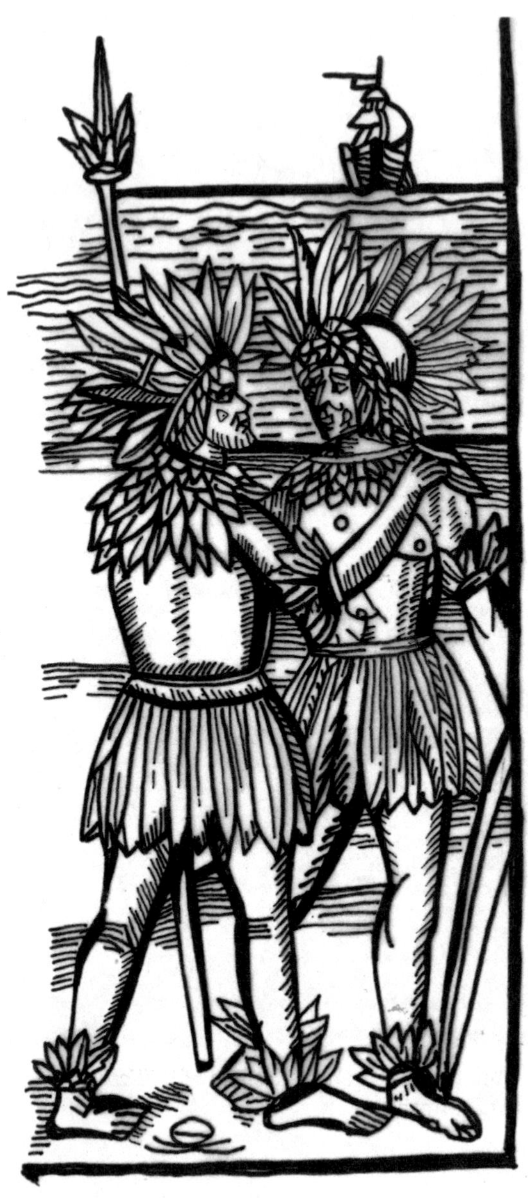

Figure 12. "The people and island that have been discovered . . ." Johann Froschauer, woodcut, Germany c. 1505–6.

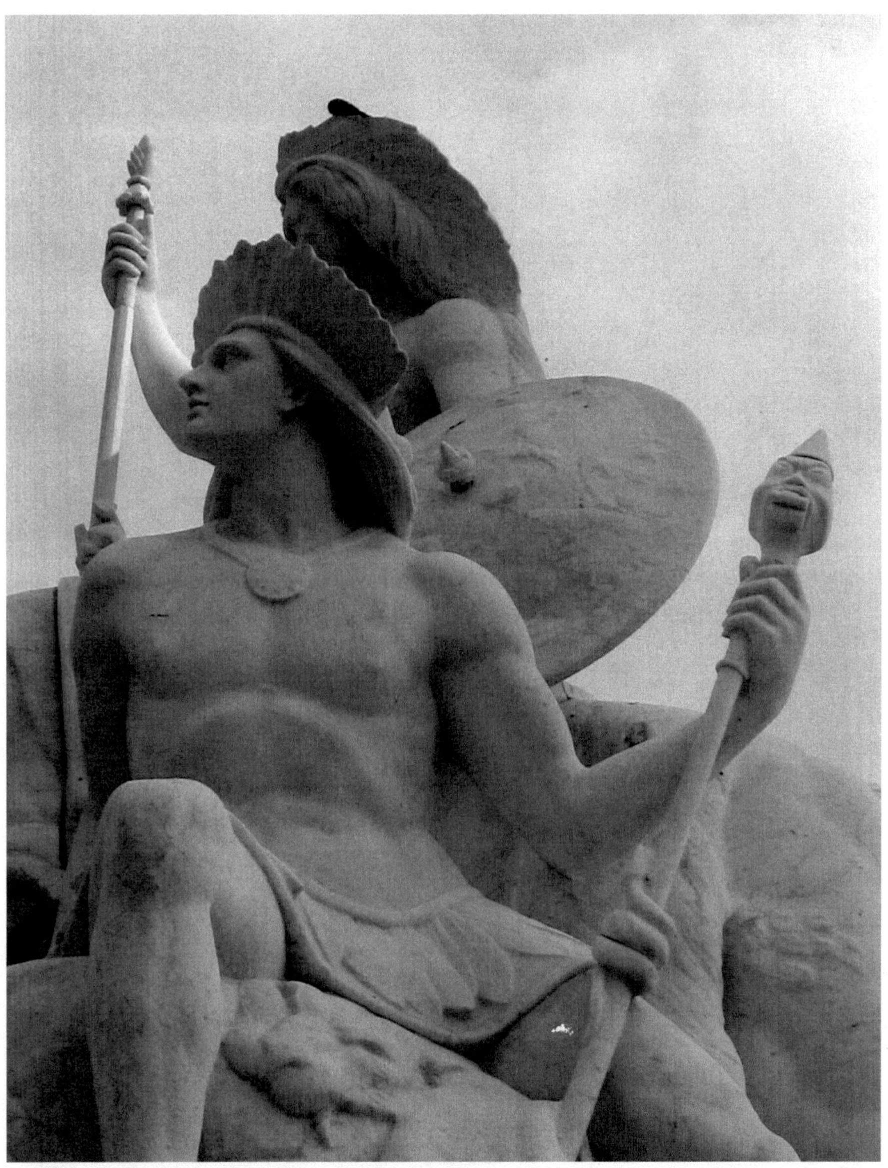

Figure 13. "America" (1864) by John Bell, part of the London Royal Albert Memorial. Photo by the author.

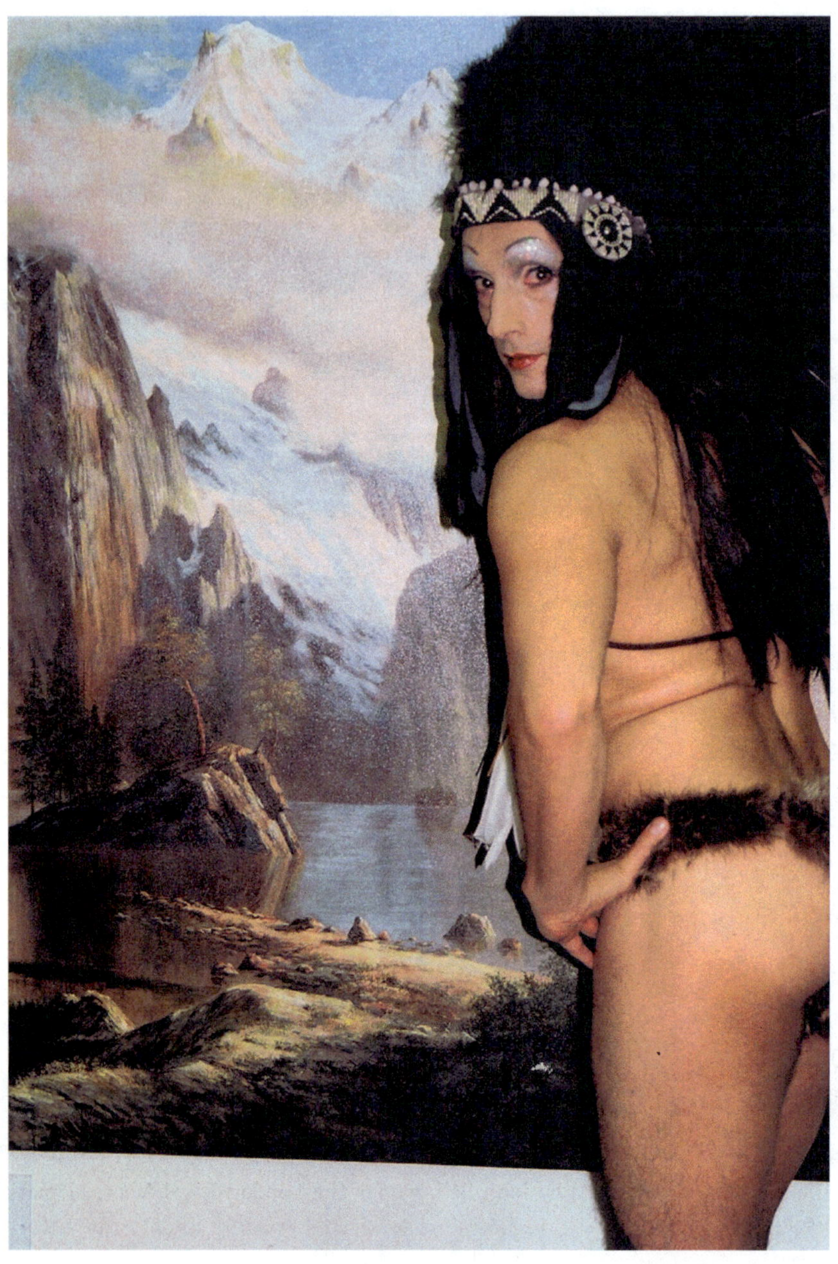

Figure 14. Cree/Metis artist Kent Monkman's alter-ego Miss Chief Eagle Testickle in front of one of his/her landscape paintings. *Credit:* Photographer Cesar Padilla (2010), used with permission.

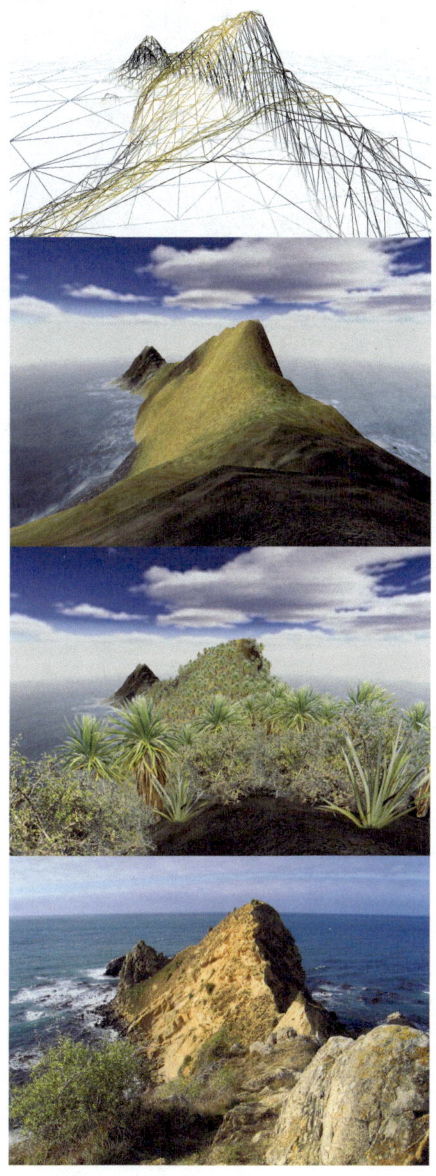

Figure 15. SimPā landscape construction of Karitane Peninsula, developed in first series of wānaka sessions.

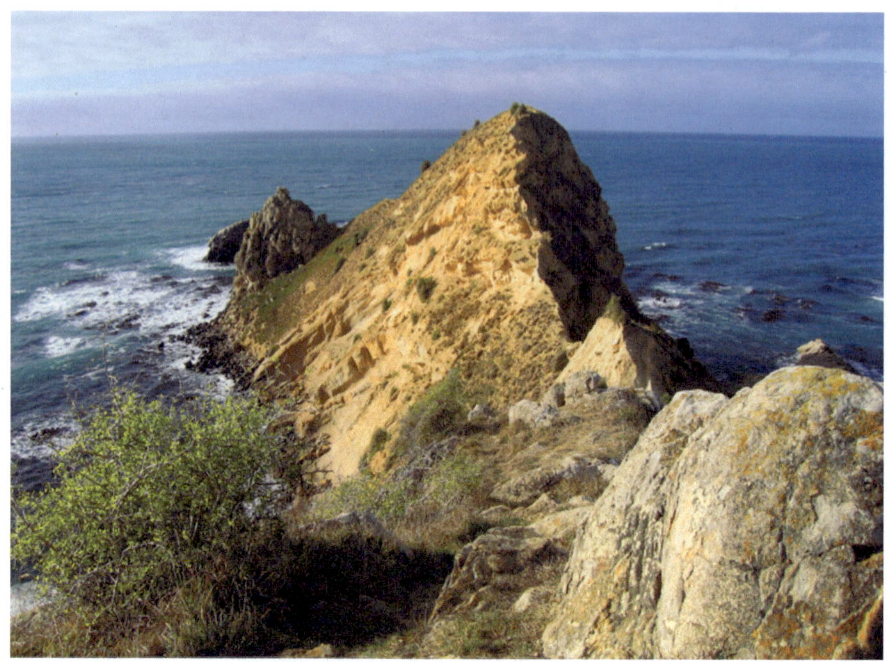

Figure 16. Karitane headland end. Photograph by Sam Mann, 2007.

V

___

DIS-EASE AND HEALING

9

Prayer with Pain

Ceremonial Suffering among the Mi'kmaq

Suzanne Owen

> We are not individuals in one sense, for, in a community of people, we are all linked together, we are all part of the same family. So, when my neighbour hurts, I hurt. I have to play a role in making sure that he stays healthy as long as he can.
>
> —Saqamaw Mi'sel Joe[1]

The introductory quotation by Saqamaw (chief) Mi'sel Joe, of the Conne River Mi'kmaq, suggests that the suffering of the individual is inseparable from the suffering of the community, and that there is a shared responsibility for the health of all individuals. There are many ways to pray amongst First Nations, but often those that involve an element of physical suffering are considered the most "powerful," and are sometimes viewed as sacrifices that can be regarded as offerings or gifts to Spirit.[2] By prayer, I refer to solitary or communal ceremonial practices for communication with other-than-human beings, to use the phrase coined by Irving Hallowell, that enhance the power of the petitioner(s).[3] It could be argued that when something is asked for—visions, healing—then something must be given in exchange in order to restore the balance and promote respectful relationships.

Many First Nations, whose members participate in traditional ceremonies that involve physical suffering, include individuals who have experienced abuse in residential schools. For them, traditional ceremonies can offer powerful forms of healing within a social setting; one form employed for such purposes is the sweat lodge.[4] Waldram, Herring, and Young have noted how "there has been an extensive revitalization of the sweat lodge as a general treatment approach for a wide variety of physical and mental health problems (as well as for social and spiritual purposes), an approach which also has the effect of reintegrating individuals into their cultures."[5] They suggest that this reintegration into the community can make ceremonies like the sweat lodge effective tools for overcoming the trauma and alienation brought on by the many and varied effects of colonization. The residential school is an institution widely recognized as a source of such trauma and alienation. One Eskasoni Mi'kmaq remarked to me that many of those who grew up in residential schools are the ones turning to "traditional spirituality." This raises questions about the link between the residential school experience and ceremonies that entail suffering of the physical body, such as the sweat lodge, fasting and, more controversially, the Sun Dance. These ceremonies, and their role in the revitalization of "traditional spirituality" among the Mi'kmaq, will be explored in this chapter

As part of my research into participation in ceremonies among First Nations, I attended the Conne River Mi'kmaq powwow, in 2003 and 2005, in order to understand the importance of "protocols"—agreed rules—for participation in ceremonies.[6] Protocols regulate a range of practices, from offering tobacco to an elder, to smoking a pipe in ceremony (e.g., a woman must refrain from doing so if "on her moon"). Although it was not the focus of my research, reactions to a particularly hot sweat-lodge ceremony I attended during the 2005 powwow led me to reflect on the place of "suffering" in the revitalization of Mi'kmaq spirituality.[7] During the ceremony, several Mi'kmaq participants from Nova Scotia and New Brunswick spoke at length about their personal suffering—domestic abuse, drug and alcohol dependency, residential school experiences, and estrangement. Two non-Mi'kmaq women left halfway through because, one told me later, it was too hot and one participant's narrative about his time in Shubenacadie Indian Residential School was "too long and harrowing." The Eskasoni Mi'kmaq sweat lodge keeper ran the ceremony as he would on the Reserve, explaining little and using it as a form of healing for himself and the other Mi'kmaq participants.

The next day, when thanking the sweat lodge keeper, I remarked that the ceremony was quite hot. He said it was not he who made the sweat hot, but the ancestors. However, a Newfoundland Mi'kmaq who did not participate in the ceremony said that "some people feel they need to suffer in the sweats."[8] This and similar comments have led me to investigate further the role of the sweat lodge and other ceremonies in transforming the effects of trauma—an aspect of the contemporary practice of traditional ceremonies that has largely gone unremarked upon by those from outside the communities involved, such as myself, yet are central for understanding the differences between indigenous performances of ceremonies such as the sweat lodge and those performed by non-indigenous people. In answering the question of why some Mi'kmaq "feel they need to suffer in the sweats," the impact of colonialism, particularly the residential school experience, becomes important for understanding the complex nature of physical suffering endured in a ceremonial context.

Although I am focussing on the Mi'kmaq of eastern Canada, the issue of suffering is common to many indigenous peoples that have been disrupted through colonialism and its institutions; therefore ways of addressing the issues have also been shared. This has had an impact on the types of ceremonies performed and ways they have been adapted to meet the needs of those who are experiencing suffering. Effective ceremonies such as the sweat lodge are now widespread among indigenous peoples in North America. The form practiced today by Mi'kmaq originated from among the Plains Indians, with some adaptations and variations.[9] While there are different attitudes among Mi'kmaq regarding suffering in the sweat lodge, a certain level of physical discomfort is to be expected. Reflecting on his role as a sweat lodge keeper, Lewis Mehl-Madrona (Cherokee-Lakota) advised: "We understand that our role is to bring the people to a point of suffering that is helpful for clearing the mind, but never taking them beyond what they can handle. My teacher said, 'If you chase the people out of the lodge, what good is that? No prayers take place.'"[10] Yet others have suggested that "one must participate a few times before the mind is able to concentrate on prayer and ignore the intense heat and discomfort."[11]

For this chapter, alongside the views of Mi'kmaq participants I met at the Conne River Mi'kmaq Powwow, I will discuss two autobiographical pieces, one by renowned Mi'kmaq poet Rita Joe (1932–2007), who attended a residential school, and the other by a counsellor, Randolph Bowers, who did not. As it is not just those individuals who have experienced residential schooling that are participating in traditional ceremonies, the wider impact of colonialism needs to be taken into consideration; for example, the creation of a distinction between "Indian" and "non-Indian" left those with mixed heritage, such as Randolph Bowers, with an ambiguous or confused status. These personal accounts show how contemporary forms of traditional spirituality can be viewed as responses to both personal and collective trauma brought on by the affects of colonization.

There is little attention given to this topic, particularly with regard to the Mi'kmaq. American anthropologist Harald Prins is an exception for including a discussion of contemporary Mi'kmaq practices in relation to colonialism and contextualizes the adoption of and reactions to Plains Indian-derived ceremonies.[12] In general, scholarship on suffering in a religious context tends to focus on rituals involving "pain," such as traditional initiation ceremonies in different indigenous cultures or mortification of the flesh in Christian or

Muslim contexts.[13] In a wide-ranging study, Ariel Glucklich explores "sacred pain" in various religious practices, including the Plains Indian Sun Dance, from a psychological-neurological perspective.[14] Glucklich offers a theoretical route toward understanding the place of suffering in contemporary Mi'kmaq spirituality.

In exploring "sacred pain," Glucklich recognizes that suffering "is not a sensation but an emotional and evaluative reaction to any number of causes," some entirely absent of physical pain, such as grief, and "that pain can be the solution to suffering, a psychological analgesic that removes anxiety, guilt, and even depression."[15] In other words, practitioners might hurt themselves physically in order to relieve psychological suffering. He elucidates the difference between "the unwanted pain of a cancer patient or victim of a car crash, and the voluntary and modulated self-hurting of a religious practitioner," which produces "cognitive-emotional changes, that affect the identity of the individual subject and her sense of belonging to a larger community or to a more fundamental state of being." Furthermore, sacred pain transforms "destructive or disintegrative suffering into a positive religious-psychological mechanism for reintegration within a more deeply valued level of reality than individual existence."[16] The turn from the self toward something greater than the self, such as the community or a "higher being," is inbuilt into many religious ceremonies. Additionally, according to Glucklich, the goal of sacred pain is "to transform the pain that causes suffering into a pain that leads to insight, meaning, and even salvation," to which I would add self-empowerment.[17] The participant emerges feeling stronger, more in control, and, in communal ceremonies, the shared experience brings participants out of a sense of isolation and alienation to a closer relationship with each other and the community.

One example of this process discussed by Glucklich is that of the Sun Dance, which he describes as "a sacrificial performance, for the good of others, for the purification of one's own community, and for the improvement of the world."[18] It is a controversial practice among the Mi'kmaq (see below). In one version of this Plains Indian ceremony, the chest is pierced by a claw, bone, or piece of wood, which is attached with a rope to a central pole. The social role of the Sun Dance is to reconcile the individual with the collective, exemplified in the experience of Aztec/Yaqui author Manny Twofeathers. During the Sun Dance, in which he experiences intense pain as his flesh is torn away from his chest, he undergoes a self-transformation "from personal to community identification."[19] Twofeathers says after being pierced, "I felt pain, but I also felt that closeness with the Creator. I felt like crying for all the people who needed my prayers."[20] The experience of intense physical pain brought about emotional counterparts, such as compassion, fear, and courage, which enabled him to overcome his alienation and feel closer to the Creator as well as his children, for whom he performed the Sun Dance.[21] As Glucklich shows, someone who engages in ceremonies that entail physical pain may undergo a transformative process, from identifying with the self to identifying with the community, and from personal suffering to compassion for others.

In trying to understand why people hurt themselves in religious contexts generally, Glucklich's study is appropriately cross-cultural, but to understand ceremonial suffering in a First Nations context specifically, it is necessary to explore accounts of First Nations' perspectives. What is not explored in Glucklich's account are the cosmological views underlying the idea among many indigenous people that the greater the voluntary suffering the greater the healing, perhaps seen as a "gift-exchange" with Spirit. One cosmological model for understanding First Nations' views about sweat lodges and fasting is suggested by Harvey Feit, who describes a reciprocal relationship between Cree hunters and the environment, using the language of gift-exchange. When the hunter has a vision of an animal before the hunt and indeed finds that animal, "that is an indicator of power."[22] Feit stresses that power is not possessed by the hunter and, since it is a gift, "the hunter is under obligation to respect that gift by reciprocating with gifts of his own."[23] The recipients are other members of the group who share in the food, while the "hunter also reciprocates to the spirits who have participated in the hunt." Thus, in traditional hunting communities, reciprocity is the basis for establishing a healthy relationship with others in the community, which includes non-human beings. "Many Cree rituals follow a similar structure," Feit comments, though he does not elaborate.[24] It may be inferred that a similar worldview underlies concepts of healing and prayer in many First Nations cultures. The language of gift-exchange can be employed to describe the actions and expectations of participants in Mi'kmaq-led ceremonies where the physical ordeal is offered as a sacrifice in order to receive power and healing.

For some, the preference for "indigenous" as opposed to other sources of ritual practice is connected to resistance to colonialism. From the 1970s, Mi'kmaq activists began to explore ways of reviving traditional spirituality as an alternative to Catholicism, the dominant religion among the Mi'kmaq. As Harald Prins observes, "many turn to their aboriginal past for cultural guidance into the future."[25] They also went out to learn ceremonies from neighboring First Nations, including the Sioux (or Lakota). Prins argues that just because a tradition is new it is no less culturally valid. One of the first ceremonies to be adopted by the Mi'kmaq was the sweat lodge. Prins was informed that the Mi'kmaq follow "the physically more challenging" "Sioux Way," which has twenty-four stones, rather than the "Cree Way," which has only twelve stones. Prins adds that the latter way is regarded "with a measure of disdain."[26] Conversely, very hot sweat lodges are commended. Anthropologist Raymond Bucko mentions in *The Lakota Ritual of the Sweat Lodge* (1998) that when someone says to a sweat lodge keeper that the ceremony was very hot it is always taken as approval. This is illustrated by an anecdote provided by a non-Indian who led a sweat lodge ceremony attended by a sceptical Lakota medicine man:

> At the end, he commented on how hot it was. He thought he would be let off the hook and it would be an easy sweat because I would not know what to do. It was what is referred to as a strong sweat.[27]

This attitude was conveyed by Mi'kmaq participants in a sweat lodge ceremony I attended during the Conne River powwow in 2003. One said that "it was really hot," which the leader took as a compliment. Two of us, both women, did not react to the extreme heat and were said to be "tough," followed by the comment that "women are tougher than men." If a sweat was "weak" (not very hot), it was often referred to as a "beginners' sweat." From these brief statements, it appears that many Mi'kmaq have favored sweat lodge ceremonies that involve greater physical suffering. According to Prins, "Traditionally, the Mi'kmaq believed that some of the most powerful 'medicine' came in visions induced by some difficult ordeal."[28] This attitude continues in the recently reconstituted Mi'kmaq traditions, including that of fasting, an intense period of prayer and abstinence from food, normally undertaken in isolation.

Fasting for several days or more is a common practice among contemporary Mi'kmaq spiritual leaders, spoken of as a first step or requirement before one can progress further in their role. One Mi'kmaq from New Brunswick, Joey Paul, told me he "fasted for his pipe" for five and a half days with the help of an elder, before receiving a pipe that would be employed in communal ceremonies, thus marking him out as a spiritual leader. An Eskasoni Mi'kmaq said he once fasted and "had a vision for a ceremony for the sea, a flesh offering, so a strong ceremony." Another Eskasoni Mi'kmaq said he returned to his own place after visiting the Cree and others and "fasted and tried to learn what the Mi'kmaw way was. This was in the eleventh year of my sobriety"—indicating the kind of journey he had been on. As a practice, fasting appears in a Mi'kmaq legend recorded by American folklorist Charles Leland in the late nineteenth century, cited by Robert Campbell, in which the Mi'kmaq cultural hero, Glooscap (or Kluskap), "through fasting and living an exemplary life, is able to bring grandmother back from death."[29] In this case, it is an offering of self-discipline in exchange for another's life. It is generally perceived that fasting is undertaken not just for the individual, but for the benefit of others in the community. In 1975, a young Mi'kmaq from Nova Scotia said he undertook a three-day fast: "I wanted to do this for myself and for my people, I think I was the only Indian to do this for a long time. It made me feel good . . . I gave my gift to the great spirit."[30] In this statement, he expresses a view that corresponds with both Glucklich's theory that these practices bring about a psychological transformation, and also the notion of "gift-exchange" highlighted in Feit's paper.

While the sweat lodge ceremony and fasting are generally accepted and still performed widely among the Mi'kmaq today, the even more "gruelling religious practice" of the Sun Dance has had a more hostile reception, often regarded as a culturally alien practice.[31] The first Sun Dance held on the Eskasoni Reserve on Cape Breton Island was led by Sioux, but most Mi'kmaq objected to it.[32] In 1988, one traditionalist woman who opposed the imported Sioux practices said:

> Do not follow their ways, they are of a very different tribe; he whoever does that abuse to their minds and bodies as well as others, must be of a violent past . . . Do you really believe that this is truly our past also, if so then we, your people are very disappointed in our forefathers . . . I am not trying to condemn you, but I do not care for their inhumane ways.[33]

She objected to the Sun Dance and other Plains Indian ceremonies not only because they were imported but because they were a form of "abuse," stemming from violence. Nevertheless, violence has been a part of the Mi'kmaq experience, too, historically through colonialism and then since the establishment of the reserve system.

"Suffering" is both something individual Mi'kmaq might be carrying within themselves, in terms of bodily, psychological, and social suffering, and is expressed in or mirrored by the physical ordeal that is endured in some ceremonies. First of all, there are certain moments in a sweat lodge ceremony that allow for the sharing of communal and personal suffering. The sweat lodge traditionally has four rounds, representing four areas of life or society. For example, in the first Mi'kmaq sweat lodge I attended, in 2003, in the second round we were asked to pray for those who were lost through drugs and alcohol. The prayers focused on members of the community—children, friends, and family who were ill or needing help. In the final round, healing for the self was sought and thanks given for the good that had already been received; surviving the residential school experience was mentioned on a number of occasions.

Shubenacadie, open from 1930 until 1967 in Nova Scotia, was the only Indian Residential School in Atlantic Canada. Ironically, Shubenacadie is a Mi'kmaw word—"place where wild potatoes grow"—although speaking Mi'kmaw was forbidden at the school. It was a place of lost childhoods, abuse, abandonment, and estrangement from the community. "Not only were Aboriginal communities distressed by the separation of children from their communities, native language and heritage, but the tragedy was worsened by the existence of physical and sexual abuse that has become evident in some parts of the sordid system."[34] Several Mi'kmaq I spoke to referred to their experience of residential schooling as a central part of their journey toward becoming a spiritual leader, carrying the pipe for the people. Joey Paul of New Brunswick, who said he was one of the first to bring the sweat lodge to the Maritimes, explained in an interview that as a child he was "kidnapped" and taken to a residential school. Back on the Reserve six years later, "there was prejudice against me when I returned, so I turned to nature, animals. Other kids mocked me for not speaking Mi'kmaw." He said he was born to become a spiritual leader. "No knowledge can make you one. You have to go and fast."[35] In other words, it is gained through physical effort, not learning. It is interesting to note that it was the rupture caused by his boarding school experience which prompted him to look to different cultures and their traditions. Disconnected from his own community, he sought guidance from other First Nations, including the Sioux, Cree, and Mohawk, bringing back the sweat lodge and other ceremonies to the Mi'kmaq.

Mi'kmaq poet Rita Joe spent four years of her childhood in Shubenacadie Indian Residential School, an experience she examines in her autobiography, *Song of Rita Joe* (1996). Born in 1932 in Cape Breton and orphaned at five, she once hid from an Indian agent and lived in different homes until age twelve when, frightened of the drinking among the adults at home, she decided to enter Shubenacadie voluntarily. Years later, Rita and her husband, who had also attended the school, went to visit one of the nuns who had taught them. Her husband vented his anger at the nun, and then said, "Sister, I don't hate you . . . It's just that I was hurt so much when I was in there, and Rita was hurt and a lot of us were hurt."[36] "For a whole year after I first came out of residential school," Rita explains, "I never went near a church. I misplaced the anger I felt about the regimentation of spiritual life in that school."[37] Just before the building was knocked down, they went to see it. Rita's husband recalled the spirits of dead children in the building, leading Rita to reflect, "Today, I think of spirits that appear anywhere on this earth as being the result of trauma . . . Over the years, so much trauma had happened in the residential school—so many people were hurt—that it played itself over and over again through the spirits."[38] Her own personal experience of trauma led her to an awareness of all those who suffered in the residential school.

Rita's first sweat lodge ceremony, led by Donna Augustine, was very hot. It was a women's sweat—"more powerful" than the men's, she says, without further explanation.[39] She also emphasizes the importance of this ceremony for prayer:

> There were times, when I was a little girl, when I prayed to be
> delivered from whatever misery I was encountering, and it didn't

happen. The misery went on and on. And then you have the unhappy realization that religion doesn't always come across for you. But prayer does help. It is possible to receive an answer. Often, when you are in the sweatlodge and you are praying, you get an immediate thought, an answer, right inside your head.[40]

Referring to the return of the sweatlodge ceremony and other traditions, Rita acknowledges that some "are still afraid of it—our brainwashing has been thorough." Despite this, the "Sacred Pipe is still being used, the sweatlodge continues—in all of this, it is the cleansing of the mind and spirit that remains uppermost in my people's minds."[41] The healing process involves, in her view, a reclamation of traditional practices and values.

Another form of suffering experienced by many Mi'kmaq is alienation. In a piece he wrote in 2008, Randolph Bowers, a trained counsellor, says he returned home from Australia to investigate his roots and Mi'kmaq approaches to healing. He felt he had an "isolated and alienated childhood" in Nova Scotia and "faced much abuse at school."

> Looking back from a professional educator's perspective, it is difficult to conclude otherwise but that the people around me then were prejudicial and unable to get beyond their own limited beliefs and values. But in the long run, these experiences made me stronger. As the years went on I developed a keen sense of compassion and empathy for people in pain.[42]

In Australia, aboriginal elders challenged him to reconnect with and learn from his own people, the Mi'kmaq, a heritage he had not really acknowledged. "This identity confusion caused me much pain. But what I realised during vision quest down under was that my Ancestors also carried pain."[43]

His identity confusion stems from being a non-status Indian—his father's side were of Mi'kmaq and Acadian French descent and had all but lost any connection to their indigenous heritage after "generations of shame and denial."[44] He writes:

> But to heal from those years of alienation, my identity needed to grow strong in other ways—by seeking solitude I found my path in life . . . By facing my hurt and confusion, something told me that there will always be a new day . . .[45]

The Mi'kmaq spiritual traditions he found back home enabled him to find healing. "In facing myself, my history and heritage was brought into the light of Mi'kmaq prayer and ceremony."[46] He gives thanks for his early spiritual awakenings that allowed him to "come into a body-awareness"; then at this point, his thoughts turn to others:

> My spirit prays with great concern today. Our children and youth are forced into such harsh circumstances in our cities and our violent communities before their spirits have time to gain strength and to awaken.[47]

Bowers considers himself part of the "lost generation." Although he himself did not go to residential school, many of his cousins did. He refers to "trans-generational trauma," sometimes specifically referring to the residential school era.[48] He thinks they each need to play a part in "healing our bloodlines."

> Yes many people have allowed these traumas to define their lives. Many have lived their lives as victims—myself included. But many of us also find our strength once again . . .[49]

After exploring his own healing process and the methods he has learned, Bowers provides insights for a cultural form of "counselling and healing based in the spirituality and spiritual ceremony and practices of Mi'kmaw People" that includes an "awareness and appreciation of the challenges still faced today that are based in the history of colonisation, oppression, and trauma faced by Mi'kmaw People" and "understanding of the impact of the residential schools on Mi'kmaw and other native communities, and of the healing work that is ahead."[50] He emphasizes the importance of ceremony in this work.[51]

As well as residential schooling, recovery from drugs and alcohol abuse is often mentioned as a reason for turning to traditional spirituality. Eleanor Alwyn, in her research on the Conne River Mi'kmaq, discusses the effect of alcoholism on families and the community and the use of ceremonies to aid healing. "They have experienced the ability to heal from what one man called 'the wobbly road' through the sweat lodge and other ceremonies, but mainly through the support of their community."[52] The importance of community in the healing of the individual is highlighted. During the 2003 Conne River pow-wow, often Newfoundland Mi'kmaq said they were motivated to heal themselves for the sake of the children of the community. As Mi'kmaq heal their past trauma and present suffering, they strengthen their future as a community.

These examples show that for many indigenous people the concept of healing is intimately intertwined with the community, and this will be explored in a different indigenous context in Chapter 10. According to sociologist Geoffrey Mercer, "the Aboriginal definition of 'health' extends beyond medically defined health outcomes to highlight physical, mental, emotional and spiritual well-being [which is] located in traditional culture and spirituality."[53] He discusses indigenous concepts of health as being inseparable from family, community, and the world, represented as a circle. Mercer quotes Joan Feather, who says:

> The circle (or wheel) embodies the notion of health as harmony or balance in all aspects of one's life . . . [Human beings] must be in balance with [their] physical and social environments . . . in order to live and grow. Imbalance can threaten the conditions that enable the person . . . to reach his or her full potential as a human being.[54]

This concept of community is also found among Mi'kmaq. "The circle is found everywhere—unity," stated one Eskasoni Mi'kmaq, explaining that he observed this among the Cree, Sioux, and other Nations.[55] The concept of the circle or wheel as representing harmony and balance has become pervasive among First Nations, especially among those who have incorporated Plains Indian ceremonies such as the sweat lodge and the Sun Dance, both of which are circular in structure. This is also noted by Glen McCabe when describing the use of the "medicine wheel" or circle, representing "balance" and "harmony," as a tool in counselling.[56]

Additionally, the community healing described by Mercer involves a "search for causes well beyond individual circumstances."[57] Referring to the Royal Commission on Aboriginal Peoples of 1996, Mercer alludes to the impact of centuries of colonialism:

> Healing, in Aboriginal terms, refers to personal and societal recovery from lasting effects of oppression and systematic racism experienced over generations. Many Aboriginal people are suffering not simply from specific diseases and social problems, but also from a depression of spirit resulting from 200 or more years damage to their cultures, languages, identities and self-respect.[58]

Concurring with this view, McCabe highlights the healing role of the sweat lodge:

> The power of the sweat lodge is a symbol of cultural integrity for Aboriginal people and serves as a reminder of the value and beauty of the traditional ways, which, in turn, encourages belief in self and community and creates hope for the future. These are two very important factors in overcoming the problems brought on by colonization and oppression.[59]

It is clear that indigenous healing involves more than the individual and, therefore, that collective forms of healing are likely to be more effective as ways to recover a sense of community. Ceremonies provide a context for this to take place. Also, there is not one indigenous healing method, but rather many that are Cree, Mi'kmaq, etc., as well as particular methods for urban, reserve, and other contexts, as the healing needs to be rooted in community in a way that addresses that community's heritage, traditions, and needs.

In the Mi'kmaq accounts discussed here, suffering is endured in a ceremonial context as a form of gift-exchange—offered in return for individual and community healing—and as a tool for transformation in order to recover from oppression, abuse, and alienation resulting from colonial methods that created social divisions, including forced relocations onto reserves, the separation of children from their families, and the distinction between status and non-status Mi'kmaq. Ariel Glucklich recognizes that voluntary self-hurting in a religious context can act as a psychological analgesic, replacing the sense of isolation with one of identification with something greater than the self, such as the community. Traditional ceremonies are perceived as facilitating a reciprocal relationship with Spirit that can bring about healing and transformation, enabling participants to overcome trauma resulting from residential school abuse and community or family breakdown, highlighted by several individuals at the Conne River powwow and in Rita Joe's autobiography, and enable Mi'kmaq, including those without Indian status such as Randolph Bowers, to come to terms with their identity and heritage within a supportive structure where more appropriate relationships are modelled as a means of remediating painful ones.

Enduring an intense physical ordeal during a ceremony can be a way of expressing or expunging past suffering. Although this might make this activity appear to be a culturally acceptable form of self-harming, its beneficial effects are far reaching and all of those who spoke about their experiences of these forms of prayer say they have gained strength and healing from them. Prayer with pain in a ceremonial context, linking the individual to community and its traditions, can offer a way to transform personal suffering into the empowerment gained through a shared healing experience, undertaken for the greater good of the people.

Notes

1. Raoul R. Andersen, John K. Crellin, and Mi'sel Joe, "Revitalization of a Mi'kmaq Community." In *Indigenous Religions: A Companion*, ed. Graham Harvey (London; New York: Cassell, 2000), 250.

2. "Spirit" (nisgam, in the Mi'kmaw language) is widely used by Mi'kmaq when speaking English to refer to the source or Creator.

3. A. Irving Hallowell, "Ojibwa ontology, behaviour, and world view" in *Readings in Indigenous Religions*, ed. Graham Harvey (London; New York: Continuum, 2002 [1960]), 17–49.

4. For research on the efficacy of the sweat lodge for aboriginal well-being, see Jeannette Wagemakers Schiff and Kerrie Moore, "The impact of the sweat lodge ceremony on dimensions of well-being," *American Indian and Alaska Native Mental Health Research: The Journal of the National Center* 13, 3 (2006): 48–69.

5. James B. Waldram, D. Ann Herring, and T. Kue Young, *Aboriginal Health in Canada: Historical, Cultural, and Epidemiological Perspectives* (Toronto: University of Toronto Press, 1995), 207.

6. This was later published as chapter 5, "Intertribal Borrowing of Ceremonies among the Mi'kmaq of Newfoundland" in Suzanne Owen, *The Appropriation of Native American Spirituality* (London; New York: Continuum, 2008).

7. This sweatlodge ceremony is mentioned briefly in Owen, *Appropriation*, 127.

8. Unlike Mi'kmaq in the rest of Atlantic Canada, Newfoundland Mi'kmaq were not taken into Indian residential schools because, prior to 1949, Newfoundland was not part of Canada. Then after confederation it was several decades before Newfoundland Mi'kmaq were able to obtain "Indian Status" (see Owen, *Appropriation*, 116–119).

9. See Owen, *Appropriation*, 113, 121–2, 127.

10. Lewis Mehl-Madrona, "Sweat Lodge," *The Journal of Alternative and Complementary Medicine* 16, 6 (2010): 609.

11. Waldram, Herring, and Young, *Aboriginal Health in Canada*, 111.

12. Harald E. L. Prins, *The Mi'kmaq: Resistance, Accommodation, and Cultural Survival* (Fort Worth, TX: Harcourt Brace College Publishers, 1996).

13. A number of studies explore the use of pain in religion, particularly in martyrdom and asceticism, influenced by the language analysis of Elaine Scarry, *The Body in Pain: The Making and Unmaking of the World* (New York: Oxford University Press, 1985)—see Ariel Glucklich, *Sacred Pain: Hurting the Body for the Sake of the Soul* (Oxford; New York: Oxford University Press, 2001), 42–44, 221 fn 7—and more recently a paper by Chris Shilling and Philip Mellor, "Saved from pain or saved through pain? Modernity, instrumentalization and the religious use of pain as body technique," *European Journal of Social Theory* 13, 4 (2010): 521–537, but few address it within a Native American or First Nations context, except in the context of "initiation" rituals or "rites of passage." In his chapter on "Emotions of Passage," Ariel Glucklich draws on Mircea Eliade, *Rites and Symbols of Initiation: The Mysteries of Birth and Rebirth* (New York: Harper Torchbooks, 1958), and Victor Turner, *The Forest of Symbols: Aspects of Ndembu Ritual* (Ithica, NY: Cornell University Press, 1967), among others, and classes the Plains Indian Sun Dance as "initiation" (Glucklich, Sacred Pain, 143), although many adults perform it annually for a number of years.

14. Glucklich, *Sacred Pain*, 10. Like Glucklich, Steven Brena uses a psychophysical approach analyzing religious discourses of chronic pain sufferers in his *Pain and Religion: A Psychophysiological Study* (Springfield, Illinois: Charles C. Thomas, 1972).

15. Glucklich, *Sacred Pain*, 11.

16. Ibid., 6.

17. Ibid., 40.

18. Ibid., 144.

19. Ibid., 146–7.

20. Quoted in Glucklich, *Sacred Pain*, 148.

21. Glucklich, *Sacred Pain*, 149.

22. Harvey A. Feit, "Hunting and the Quest for Power: The James Bay Cree and Whitemen in the 20th Century," in *Native Peoples: The Canadian Experience*, ed. R. Bruce Morrison and C. Roderick Wilson (McCelland & Stewart, 2nd ed., 1995), Part I: The Contemporary Cree Hunting Culture, http://arcticcircle.uconn.edu/HistoryCulture/Cree/Feit1/feit1.html.

23. Feit, "Hunting and the Quest for Power," Part I.

24. Ibid.

25. Harald E. L. Prins, "Neo-Traditions in Native Communities: Sweat lodge and Sun Dance Among the Micmac Today," in *Actes du Vingt-Cinquième Congrès des Algonquinistes*, ed. William Cowan (Ottawa: Carleton University Press, 1994), 385, 392.

26. Prins, "Neo-Traditions in Native Communities," 390.

27. Raymond Bucko, *The Lakota Ritual of the Sweat Lodge: History and Contemporary Practice* (Lincoln, NE: University of Nebraska Press, 1998), 228.

28. Prins, *The Mi'kmaq*, 72.

29. Robert A. Campbell, "Bridging Sacred Canopies: Mi'kmaq Spirituality and Catholicism," *The Canadian Journal of Native Studies* 18, 2 (1998): 306.

30. Quoted in Prins, "Neo-Traditions in Native Communities," 387.

31. Prins, "Neo-Traditions in Native Communities," 390.

32. Ibid., 391.

33. Quoted in Prins, "Neo-Traditions in Native Communities," 392.

34. Martin Thornton, "Aspects of the History of Aboriginal People in their Relationship with Colonial, National and Provincial Governments in Canada," in *Aboriginal People and Other Canadians: Shaping New Relationships*, ed. Martin Thornton and Roy Todd (Ottawa: University of Ottawa Press, 2001), 13. There are a number of studies on abuses that took place in the residential school system in Canada, such as Assembly of First Nations, *Breaking the Silence: An Interpretive Study of Residential School Impact and Healing as Illustrated by the Stories of First Nations Individuals* (Ottawa: Assembly of First Nations, 1994) and J. R. Miller's historical study of residential schools, *Shingwauk's Vision: A History of Native Residential Schools* (Toronto: University of Toronto Press, 1996). See also Mary-Ellen Kelm, *Colonizing Bodies: Aboriginal Health and Healing in British Columbia 1900–1950* (Vancouver: University of British Columbia Press, 1998). For a list of others, see Jean Barman, "Aboriginal Education at the Crossroads: The Legacy of Residential Schools and the Way Ahead." In *Visions of the Heart: Canadian Aboriginal Issues*, eds. David Allan Long and Olive Patricia Dickason (Toronto: Harcourt Brace & Company, 1996), 271–303. For a Mi'kmaq account of the impact of colonization, including residential schooling, see Daniel N. Paul, *We Were Not The Savages: Collision between European and Native American Civilizations* (Halifax, NS: Fernwood Publishing, 3rd ed., 2006).

35. Owen, *Appropriation*, 123, 140.

36. Rita Joe, with Lynn Henry, *Song of Rita Joe: Autobiography of a Mi'kmaq Poet* (Lincoln: University of Nebraska Press, 1996), 48.

37. Joe, *Song of Rita Joe*, 56.

38. Ibid., 57.

39. Ibid., 146–7.

40. Ibid., 151.

41. Ibid., 154.

42. Randolph Bowers, *Reconnecting with the Mi'kmaq* (Mi'kmaq Resource Centre Special Collection: University of Cape Breton, 2008), 10. http://mikmawey.uccb.ns.ca/Rsrchrprt_Rconnctng_bowers_08.pdf

43. Bowers, *Reconnecting with the Mi'kmaq*, 11.

44. Ibid., 29.

45. Ibid., 19.

46. Ibid., 18.

47. Ibid., 19.

48. Ibid., 32, 36.

49. Ibid., 31, 32.

50. Ibid., 26.

51. Ibid., 27.

52. Eleanor Alwyn, "Circletalk: If We Don't Know Where We Come From, We Don't Know Where We're Going" (paper presented at the Interinstitutional Consortium for Indigenous Knowledge conference, 2004). http://www.ed.psu.edu/icik/2004ConferenceProceedings.html.

53. Geoffrey Mercer, "Aboriginal Peoples: Health and Healing" in *Aboriginal People and Other Canadians: Shaping New Relationships*, eds. Martin Thornton and Roy Todd (Ottawa: University of Ottawa Press, 2001), 144.

54. Quoted in Mercer, "Aboriginal Peoples," 144.

55. Owen, *Appropriation*, 125.

56. Glen McCabe, "Mind, body, emotions and spirit: reaching to the ancestors for healing," *Counselling Psychology Quarterly* 21, 2 (2008): 145.

57. Mercer, "Aboriginal Peoples," 144.

58. Mercer, "Aboriginal Peoples," 156.

59. McCabe, "Mind, body, emotions and spirit," 158.

10

———————

Coping with Colonization

Aboriginal Diabetes on Manitoulin Island

Darrel Manitowabi and Marion Maar

Introduction

Diabetes is currently emerging as a global health issue, but Aboriginal people in Canada have suffered from epidemic rates of type 2 diabetes and its many medical complications for several decades.[1] Biomedical studies of this phenomenon among First Nation peoples have predominantly focused on physical risk factors, such as genetics, nutrition, and exercise, in the prevention and management of Aboriginal diabetes.[2] In contrast, Aboriginal worldviews perceive health and well-being as balancing the physical, mental, emotional, and spiritual aspects of the body.[3] As a consequence, many Aboriginal people see a strong connection between the occurrence of diabetes in First Nations and the loss of spirituality and overall community wellness, which they believe can be linked to colonial histories, power imbalances, stress, racism, and intergenerational trauma.[4]

Research examining Aboriginal understandings of the pathways that lead to this disease are clearly necessary. In this chapter, we examine Aboriginal perspectives related to coping with the experience of diabetes using a holistic approach focused on the role of culture, spirituality, and emotional wellness revealed in narratives of eighteen Aboriginal people living with diabetes on Manitoulin Island, in north-central Ontario. As "living well" is synonymous with health and wellness for Manitoulin Island Aboriginal peoples, we suggest that addressing the legacy of colonization is an important determinant in type 2 diabetes management.

Background

In 2006, Manitoulin Island First Nations in north-central Ontario initiated a comprehensive diabetes care and prevention research program that pointed to differences in coping strategies between Aboriginal and non-Aboriginal people.[5] Although many Aboriginal patients manage well, a high percentage struggle, and over-all health outcomes are generally poorer among Aboriginal patients with diabetes.[6] Within this context, this chapter explores the impact of spiritual, emotional, and social wellness on the development of diabetes and secondary complications among Manitoulin Island Aboriginal peoples, using in-depth interviews with eighteen participants.

Type 2 diabetes mellitus is an extremely serious health problem not only on Manitoulin Island, but within many Aboriginal populations in North America.[7] Women living on Canadian reserves have a more than fivefold risk of death from diabetes compared to the general Canadian population[8] and the prevalence of diabetes is about 3.6 and 5.3 times higher among First Nations men and women compared with the general Canadian population.[9] On Manitoulin Island, the prevalence of diabetes varies in communities, ranging from approximately 7 to 30 percent of adults being affected by this disease. Type 2 diabetes is also observed in an increasing number of Aboriginal children and adolescents.[10] This is alarming, since these children are likely to develop the complications of diabetes at a younger age, with detrimental long-term health consequences throughout their adult lives

Several studies have focused on the contrast between Aboriginal and biomedical explanatory models of the causation of diabetes.[11] Generally these studies found that Aboriginal explanatory models identify stress, Aboriginal susceptibility, and a loss of traditional lifestyle as causal agents. Biomedical causation focuses on obesity and genetic susceptibility; the management of diabetes mainly addresses obesity, with a focus on dietary restrictions and exercise plans. The implication of the Aboriginal model, however, is that an understanding of culturally specific explanatory models is necessary in order for health providers to deliver culturally competent care for Aboriginal patients.

Within a particular culture, certain coping strategies are more effective than others in dealing with stressors and improved well-being.[12] Approaches to coping include: problem-focused coping, which focuses on addressing the problem; emotion-focused coping, which focuses on improving negative emotions; meaning-focused coping, which addresses the meaning of the problem; and social coping, which focuses on social support to address the problem.[13] Each of these coping styles can result in positive or negative results for different patients. A better understanding of common coping strategies within a culture can improve the relationship between health providers and patients, thus improving patient outcomes. With the exception of the work by Iwasaki et al.,[14] the coping strategies Aboriginal people employ to manage diabetes and their implications for holistic care provision have not been systematically investigated.[15]

Colonization processes impact dramatically on determinants of health, and Aboriginal health studies must take this into consideration. In Canada, colonization resulted in Aboriginal displacement, marginalized land bases, sociocultural disruption, assimilation, external political control, state dependency, economic encapsulation, low-level social services, and the imposition of racial hierarchies situating Aboriginal peoples below non-Aboriginal peoples.[16] Mary-Ellen Kelm calls this process "colonizing bodies" and it is characterized by colonially induced poverty, undernourishment and poor housing all leading to a susceptibility to ill-health.[17] Naomi Adelson points out these Aboriginal health disparities in turn reveal an embodiment of inequality in Canada, and that while genetics plays a role in chronic diseases such as diabetes, there is a need to "examine the role of changing diets, changing or limited work options, poverty, access to resources, societal stressors, and the cultural variations of foodstuffs as part of the more complex picture of disease in the contemporary context."[18] In short, there is an urgent need to pay attention to the deep and enduring legacies of colonization, and the profound impacts they have on Aboriginal lives, when seeking to understand both the causes and the most effective way to approach the treatment of First Nation patients with type 2 diabetes.

Aboriginal Diabetes on Manitoulin Island

Manitoulin Island is a large island located in northern Lake Huron in north-central Ontario. The island is home to approximately 5,000 Aboriginal peoples of Ojibwa, Odawa, and Potawatomi descent living in seven First Nations communities. These First Nations include: Sheguiandah, Whitefish River; Aundeck Omni Kaning; M'Chigeeng; Sheshegwaning; and Zhiibaahaasing, which are connected under the United Chiefs and Councils of Manitoulin (UCCM), and Wikwemikong.[19]

In 2005, the Health Boards in the UCCM territory invited one of the authors (Maar) to conduct consultations with local communities and organizations to determine diabetes-related research priorities. This process provided the foundation for an interdisciplinary diabetes care and prevention research program. The research consultation employed a participatory action approach and was completed in 2006. Based on 22 focus groups conducted with Elders, paraprofessionals, specialized providers, mental health care providers, primary care providers, and patients, there was broad consensus that research is necessary to identify cultural, emotional, and spiritual aspects of diabetes that are part of the Aboriginal worldview, and their respective influence on patients' coping strategies.

Results also showed that health care workers on Manitoulin Island were in agreement that they see, on average, poorer outcomes for Aboriginal peoples with diabetes than they would see in their non-Aboriginal patients. While they acknowledged that some of their Aboriginal patients are doing well managing their illness, many Aboriginal patients are struggling. Those who are struggling generally show evidence of poor glycemic control and high rates of secondary complications such as heart disease, kidney failure, blindness, and amputations.[20]

Many health care providers note the need to address more than just the physical aspects of the disease in order to help patients cope with diabetes. The Maar et al. study contends that health care providers need to understand more fully what conditions in a patient's life are linked to good coping skills.[21] In particular, the link to spirituality, emotional health, family history, and exposure to intergenerational trauma were seen as important areas to explore. For those patients who are coping very well, health care providers suggested that research should include the collection of qualitative data regarding their support system, their connection to Aboriginal culture and spirituality, and to the community.

Methods

In 2007, we initiated a participatory action research project focused on the above needs, in close collaboration with the Manitoulin Island First Nation stakeholders. This research included the following two components: 1) to conduct a community-based workshop addressing how Aboriginal spiritual health can be researched respectfully, and 2) to collect pilot data to develop a framework for coping mechanisms that Aboriginal peoples use to manage diabetes.

A community-based workshop, with participation from Manitoulin Island First Nations people and health care providers, provided the foundation for this project. Discussions held during this workshop revealed that in order for respectful diabetes research to take place, it must be guided by the local Aboriginal concept of the Seven Grandfather Teachings that include wisdom, love, respect, bravery, honesty, humility, and truth. In addition, researchers must encourage community control, address community needs, and include community input.[22] These goals were realized with the help of a community-based research steering committee to oversee the project, and with close collaboration with Aboriginal diabetes educators. These educators were also instrumental in recruiting patients to voluntarily participate in this project.

Working closely with the steering committee, we formulated interview questions and developed a sampling strategy. We determined that it was important to incorporate proportional representation of First Nations reflective of the population (for example, the smallest community population is 35, while the largest is 872), and to include an equal representation of male and female participants.[23] Furthermore, we attempted to recruit an equal number of patients who, in the view of community diabetes educators, coped positively with the management of diabetes, versus those who struggled to cope. In total, eighteen interviews were completed, audio-recorded, transcribed, and analyzed thematically.

Results

Several themes were invoked in patient narratives, including the historical context of diabetes, causation, coping strategies, the problem of alcohol in diabetes management, the concern for the youth, and solutions for diabetes prevention and management. In the following, selected narrative profiles reveal the meaning of these issues.

A consistent theme evoked in patient narratives is the historical context of diabetes and its emergence alongside the changing character of First Nations communities. One patient remarked:

> Growing up was great, actually. We always had lots to do. We're talking about the late 60s early 70s: I was just going into my teens. Back then there weren't that many homes here. Probably about a quarter of what we have here now and a lot of this area was all bush. So we were always out and we ate a little better. I don't know if we ate better, here I am trying to think back on it, we ate a lot of macaroni and things like that, pastas. I don't know if that was actually all that good for us. We did have a lot of wild game, natural foods—veggies, we used to grow our own. We were always out in the bush, running around. We were always involved in sports activities growing up and television was just a real treat because we grew up with only about two stations. We didn't have very much to watch so we were out quite a bit. We were just active. It's a complete reversal of what I see today. Kids always say there's nothing to do but we used to be able to do without the things that they have today.

It is apparent from this narrative that regular physical activity and healthy foods were a part of the lived experience in the 1960s and 1970s in this patient's community, and there was a general sense of wellness. The interviewee suggests the experience of the youth in present times is a reversal of the past.

Causation

Many patients mentioned that diabetes was not an issue while they were growing up and it is a relatively new phenomenon. In terms of causation, many stated that genetics was a possible influence, however many stressed that diet, weight gain, stress, and alcohol were factors triggering the onset of type 2 diabetes. The next quote provides insight:

> (It is caused by) [j]ust lifestyle more than anything. It was probably genetics—my grandparents [had it]. But a lot of it was just my lifestyle. I used to be a heavy drinker. I ate a lot of snack foods, a lot of fast foods; probably drinking almost a litre of pop everyday didn't help. For me, it was probably a combination of all that, just the lifestyle, not doing anything that much or very little. I wouldn't say that I didn't do anything. And what I ate—that kind of lifestyle probably contributed to it the most. Drinking everyday, things like that. And my diet—or lack of a diet.

In the above narrative the patient acknowledges a possible genetic predisposition, but stresses that diabetes is linked to lifestyle, diet, and alcohol consumption.

Coping Strategies

In terms of diabetes management, although most felt they managed fairly well, common challenges were nevertheless diet, physical activity, and the affordability of healthy food. With respect to coping, patients remarked that family members and friends are supportive, diabetes educators are helpful, and others simply have a positive outlook on life and are determined to deal with diabetes. For instance, one patient stated:

> . . . yes as long as I've had it. . . . I learned to cope with it. Let's put it that way. I've learned to cope with diabetes and the hardships to go along with it. . . . I have diabetes, it's easier to cope with from a day to day basis. That's one thing I learned to do . . . enjoy my life with diabetes. You learn to cope with it. So it's part of your life and always will be a part of your life. You got to have coping mechanisms to help you heal with . . . like I told you about my family . . . you know . . . a lot of people

die . . . they don't have diabetes but they die from cancer and things like that you know and . . . sometimes when I'm down and out about myself I think of other people that are worse off than I am and that helps me cope with anything . . . it helps me cope better. . . . It all depends on you. If you think negatively all the time "Oh I don't feel good . . . I'm never going to have a life" stuff like that, well . . . it'll probably happen if you don't take it. You got to take it . . . you have to. Like I say there's always some kind of a coping mechanism to relieve whatever you have.

On the other hand, there are those who are struggling deeply with the progressive nature of diabetes; for instance, one patient in this position stated:

. . . it's very difficult because you try, you try everything . . . um then again as you go through the stages you just keep, you just never get better and you just get worse . . . I think my blood sugars have improved over the years . . . and then I'm at the point where I'm going downhill. I try really hard to stay on track, but as I go through the different stages and the complications of diabetes, it's kind of hard some days eh, like you just wish the pain would go away and you could get better or if you can have some kind of a magic solution to get over it.

These narratives reveal the various realities in struggling to live with diabetes. Some participants were at advanced stages of diabetes complications. One had severe kidney problems, while another had challenges with controlling blood sugar levels. Though a positive or resilient mental attitude can nurture an acceptance of diabetes, the reality is that the lifestyle changes required from the medical perspective are difficult to sustain and as diabetes progresses a gradual deterioration seems inevitable.

The Problem of Alcohol

A few participants stated that alcohol abuse leads to poor diabetes management. In particular, one patient stated:

. . . the biggest one for me in all of this would be um alcoholism. That would be my own disease. So when I said that was a big cheat, and that was a challenge, it was socially a challenge because everyone around me had to learn that I had diabetes. I didn't realize the extent of the seriousness of what it could be doing to my body so. . . . So for me it's like a cycle, it was like a big cycle,

watching family members who drank and still had diabetes so to
me that was the accepted norm and I didn't, I didn't make any
changes I didn't really take any steps toward a healthier lifestyle.

In the cases where there was a connection between alcohol consumption
and poor diabetes management, chronic poverty and unemployment were
also factors. Those participants spoke about the difficulty of living on social
assistance and the daily struggle to get by.

A widespread community issue discussed by participants is the grave
concern that the health and well-being of the youth is more threatened by
diabetes than it was in previous generations. This was seen as a universal
issue, many expressing a need for community-wide interventions to prevent
future spread of diabetes. One patient stated:

I think there will be a lot more cases of diabetes in the community
within the next ten years. I can see their lifestyles (of the youth),
their attitudes, the path a lot of them are leading. I think it will
increase within the next ten years. It won't go down. I don't know
what our complete stats are here but I think it's only going to
increase. That's basically because of lifestyles and the attitude—it's
not taken seriously.

In many ways, some patients felt helpless and powerless regarding the
spread of diabetes in future generations of their community. During interviews the tone and expression generally became sombre as they reflected that
the current problem of diabetes would likely become a greater problem for
First Nations communities to face in the future.

Solutions for Overcoming Diabetes

In terms of a solution, many emphasized the need for education, better diet,
and improved physical lifestyle. Although these seem to be consistent and
even generic responses to diabetes prevention and management, some were
forthright in their assertion that a paradigm shift in the community is needed,
specifically a return to empowered life ways. This is apparent in the following quote:

Well when you think about, well I think with the traditional
ways . . . and that, the traditional way we eat . . . before diabetes
wasn't there when . . . people were living like that. I remember,
like I said eating deer and beaver, seagull eggs and stuff like
that . . . food from the land, I think we need to go back to the

food and even the garden. That kind of stuff and then we need to
go back to . . . physical work, you know having a garden. . . . we
need to go back to, to some of that, or we can be creative and
do a community garden or something. It's too easy when you go
to the drugstore everything is . . . pre-made, you know you got
meat pre-made and all you have to do is put it in the microwave,
it . . . it's just too easy. . . . I think the spiritual part, just having
that understanding about the Creator and the Creator giving you
this gift of life . . . and learning about how to take care of that
life and that's you . . . and you look at the four areas, physical,
mental, emotional, spiritual, they're all connected and I think that
needs to be reintroduced.

All of the above narratives regarding diabetes history, its causes, coping strategies, its impact on the youth, and solutions for overcoming it, provide insights
on Aboriginal perspectives on wellness and diabetes. In particular it is possible
to examine the meaning of Aboriginal wellness as a positive coping strategy
for diabetes. Of the eighteen interviewed, a few mentioned the incorporation
of Aboriginal medicinal teas as useful in diabetes management, and some
mentioned a direct connection between cultural and spiritual worldview and
following an empowered way of life as coping mechanisms. Specifically, two
of the individuals actively maintain traditional lifestyles by consulting traditional healers, attending ceremonies, spending time on the land, and having
a positive outlook on life.

It is apparent there are complex issues at play concerning Aboriginal
cultural, spiritual, and emotional wellness and the onset and management of
type 2 diabetes for Manitoulin Island First Nations. Wellness is a way of life
situated within a broader historical, social, and political context. This situation is similar to the health and wellness work among the James Bay Cree
of Quebec studied by Naomi Adelson. According to Adelson,

Although only two made direct connections to Aboriginal cultural practices, all participants remarked on the social and political aspect of diabetes
management as being significant. This is in reference to changing Aboriginal
lifeways, a genuine concern that diabetes is a community issue, and the need
for an action plan to prevent a future epidemic.

[f]or the Cree there is no word that translates into English as
"health." The most apt phrase is *miupimaatisiiun* or, as I trans-
late it, "being alive well." "Being alive well" constitutes what one
may describe as being healthy; yet it is less determined by bodily
functions than by the practices of daily living and by the balance of
human relationships intrinsic to Cree lifestyles. "Being alive well"

means that one is able to hunt, to pursue traditional activities, to
eat the right foods, and (not surprisingly, given the harsh northern
winters) to keep warm. This is above all a matter of quality of
life. That quality of life is linked, in turn, to political and social
phenomena that are as much a part of the contemporary Cree
world as are the exigencies of "being alive well."[24]

For Adelson, the context of health is important here since the Cree of James
Bay have been affected by a massive hydroelectric project, and have thus been
forced into a defined political and economic relationship with the government of Quebec and the public utility, Hydro Quebec. This economic encapsulation resulted from a "developmentally" induced altered landscape with
diverted waterways and flooded traditional territories that have affected and/
or destroyed Aboriginal hunting economies. For the Quebec Cree, "strategies
of health—*miupimaatisiiun*—connect individual bodies to that larger political
process; in this way links are formed between health, cultural assertion, and
dissent within both individual bodies and the body politic."[25] According to
Adelson, her fieldwork experience was summed up by one question posed by
an informant: "If the land is not healthy then how can we be?"[26]

The First Nations of Manitoulin Island are not affected by a hydroelectric project, but share a similar colonial history with the Cree and have a
similar word connoting health, "*mnaamodzawin.*"[27] Aboriginal lifeways such
as governance, control of local resources, and foods were altered in the colonial process and continue to be altered because of political and economic
encapsulation in modern times.[28] During the interviews, stories of poverty
and varying degrees of social suffering were recounted, such as challenging
upbringing, underemployment, alcohol abuse, and a general disempowered
way of life.

Embedded within these themes are reduced Aboriginal traditional territories resulting from treaties, in turn minimizing land bases and First Nations
political and economic power. Within this framework, it is possible to make
a connection with these narratives and Adelson's reference to a "strategy of
health" that connects "bodies to larger political processes." As stated earlier in
this chapter, a diabetes consultation as well as subsequent research has identified poorer type 2 diabetes outcomes of Aboriginal peoples in comparison to
non-Aboriginal peoples on Manitoulin Island. Specifically, our findings suggest
early onset of diabetes and related complications such as heart disease among
Aboriginal peoples appear to be a bodily manifestation of colonization.[29] Our
research therefore supports Kelm's argument that Aboriginal bodies make
these inequities visible, as evidenced by high rates of diabetes and related
complications.

Such a perspective moves beyond a purely biomedical framework of diabetes to a more holistic Aboriginal model. Jennifer Poudrier has made similar connections specific to biomedical genetic studies of diabetes among Canadian Aboriginal peoples. By critically examining the publications on the thrifty gene theory and the Sandy Lake Oji-Cree studies of northwestern Ontario, Poudrier suggests this approach produces a eugenic, neo-colonial perspective that is based on a naturalization of the political, economic and historical conditions (colonization).[30] The implication is that Aboriginal genes are the "problem" and that colonization is never accepted as the root cause of the observed high rates of type 2 diabetes. Genetic research therefore stigmatizes the people they describe and sets the basis for future race-related research.[31]

This colonial impact on Aboriginal wellness is similar to the work of the Cree social worker, Michael Hart. He employs the same concept used by Manitoulin Island Aboriginal peoples, and states that *mino-pimatisiwin* is the "good life," and reaching the good life involves the individual, family, and community.[32] A main strategy of reaching the good life involves empowered lifeways inclusive of sharing, respect, an acknowledgment of the unity of all community members, and balance of humanity with nature.[33]

The antithesis of *mino-pimatisiwin* is "domination and oppression."[34] For Aboriginal people the history of colonization has affected the Aboriginal nation, community, family, and individual. As a state of spiritual well-being, it has also affected ceremonial practices and human development.[35] Key to this sense of well-being is an Aboriginally-empowered path of life inclusive of a biological and cultural sense of being. Patient narratives revealed a need to address the physical, emotional, cultural, and spiritual aspects of life; thus echoing Hart's message.

In the context of diabetes, it is possible to envisage such an Aboriginal sense of well-being as providing an understanding of Aboriginal cultural, spiritual, and emotional wellness that is necessarily linked to broader historical, social, and political processes. Specific to the themes evoked in patient narratives, the association between coping and diabetes is influenced by life experience, family support, socioeconomic status, and individual agency. Despite the various coping experiences, there is consensus among all those interviewed that diabetes is a new disease and is a growing concern for all Manitoulin Island First Nations, particularly the youth. Furthermore, there is a desire to invest in empowered lifeways such as promoting a better diet, a renewed connection with nature, nurtured identity, and greater physical activity as ways to overcome diabetes. It is possible to draw linkages between the insights of Adelson's concept of Aboriginal well-being as a "strategy of health linked to broader processes;" Kelm's "colonizing bodies;" and Hart's "good life" of sharing, respect, and balance. Participants in this research corroborated these ideas, especially the struggle of living with diabetes, and the genuine concern for community and the future. As such, diabetes management is a localized biological engagement of Aboriginal bodies with broader power struggles.

Conclusion

In this study we examined the relationship between spiritual, emotional and social wellness and diabetes management on Manitoulin Island First Nations. Patients viewed diabetes as being a recent issue linked with loss of control over lifeways, such as active lifestyles, access to wild or homegrown foods, and a spiritual way of life. Contrasted with the present, there is a heavy reliance on manufactured foods and an overall sense of powerlessness. Participants blamed high rates of diabetes and poor management on these changes in lifestyles in First Nation communities. Thus colonization and power imbalances have led to poorer food choices, less physical activity, and generally a sense of unwellness in the community; a general loss of *mnaamodzawin*.

In contrast, the effective management of diabetes necessitates incorporating that which was lost due to colonial forces: healthier foods and regular physical activity. Reincorporating these was seen as today's challenge. Alcohol and addictions, also a consequence frequently linked to colonization, further exacerbates the poor health of some people living with diabetes. All were concerned about the increasing rates of diabetes in children and youth. These issues are consistent with the literature examining Aboriginal wellness in the context of broader historical and political processes. As such, the discourse of Aboriginal diabetes must not be strictly tied to a biomedical conception of diseased biological bodies, but must also link historical and political processes *outside* of the body as bearing relational consequences *within* the body. Despite this seemingly dire predicament, the Aboriginal peoples on Manitoulin Island express the hope that a paradigm shift can occur in First Nations, returning their control over foods and life-ways.

Regarding the broader context of Aboriginal diabetes research, Nancy Scheper-Hughes contends that there is a need to: ". . . reframe the meaning of diabetes as a socio-political pathology and to place the disease outside and beyond the body of the individual sufferer and to see it as the consequences of genocide and its aftermath and in the signature that these collective losses have left on the bodies and even the physiologies and chemistries of Indigenous Peoples in the past and today."[36] With this in mind, the issue of diabetes necessarily is inclusive of connections between individuals, families and community and politico-historical processes. From this perspective, it is possible to see diabetes-coping not as an individualized process, but as one that connects individuals, families, and communities to the past, present, and future of the Manitoulin Island Aboriginal peoples.

Acknowledgments

This research was made possible by a grant from the Social Sciences and Humanities Research Council of Canada. On Manitoulin Island, Valerie Beaudin and Eleanor Debassige provided invaluable support by arranging interviews. Cheri Corbiere and Lorrilee McGregor helped with interviews. Lastly, we are grateful to the Members of the Diabetes Research Steering Committee for their ongoing guidance and to Lisa Boesch for providing transcriptions.

Notes

1. The term "Aboriginal people" is commonly used in Canada to describe all Indigenous peoples. Specific to this chapter, we use Aboriginal to refer to the collective people and First Nations to refer to Aboriginal Reserve communities. Medical problems related to type 2 diabetes include heart and kidney disease, nerve damage, loss of vision, and amputations as well as depression and lack of energy.

2. See for instance, J. Sunday, J. Eyles, and R. Upshur, "Applying Aristotle's doctrine of causation to Aboriginal and biomedical understandings of diabetes," *Culture, Medicine and Psychiatry* 25 (2001), 63–85.

3. For an example specific to Aboriginal diabetes research see H. Nabigon, R. Hagey, S. Webster, and R. MacKay, "The Learning Circle as a Research Method: The Trickster and Windigo in Research," *Native Social Work Journal* 2 no. 1 (1999), 113–137.

4. See, M. L. Ferreira and G. C. Lang, eds., *Indigenous Peoples and Diabetes: Community Empowerment and Wellness* (Durham: Carolina Academic Press, 2005).

5. M. Maar, L. McGregor, and M. Sutherland, *Results of the United Chiefs and Councils of Manitoulin Diabetes Care and Prevention Research Consultation* (2006), Unpublished report prepared for the United Chiefs and Councils of Manitoulin Health Boards.

6. M. Maar, D. Gzik, and T. Larose, "Beyond expectations: Why do Aboriginal and Euro-Canadian patients with type 2 diabetes on a northern, rural island demonstrate better outcomes for glycemic, blood pressure and lipid management than comparison populations?" *Canadian Journal of Diabetes* 34 no. 2 (2010), 127–135.

7. T. K. Young, J. Reading, B. Elias, and J. O'Neil, "Type 2 diabetes mellitus in Canada's First Nations: status of an epidemic in progress," *Canadian Medical Association Journal* 163 no. 5 (2000), 561–566. For a global Indigenous incidence of diabetes see M. Naqshbandi, S. B. Harris, J. G. Esler, and F. Antwi-Nsiah, "Global complication rates of type 2 diabetes in Indigenous peoples: A comprehensive review," *Diabetes Research and Clinical Practice* 82 (2008), 1–17.

8. Y. Mao, B. W. Moloughney, R. M. Semenciw, and H. I. Morrison, "Indian Reserve and Registered Indian Mortality in Canada," *Canadian Journal of Public Health* 83 (1992), 350–353.

9. See for instance, H. J. Dean, T. K. Young, B. Flett, and P. Wood-Steiman, "Screening for type-2 diabetes in aboriginal children in northern Canada," *Lancet* 352 (1998), 1523; and S. B. Harris, B. A. Perkins, and E. Whalen-Brough, "Non-insulin-dependent diabetes mellitus among First Nations Children: A new entity among First

Nations people of north western Ontario," *Canadian Family Physician* 42 (1996), 869–876.

10. See for instance, H. J. Dean, T. K. Young, B. Flett, and P. Wood-Steiman, "Screening for type-2 diabetes in aboriginal children in northern Canada," *Lancet* 352 (1998), 1523; and S. B. Harris, B. A. Perkins, and E. Whalen-Brough, "Non-insulin-dependent diabetes mellitus among First Nations Children: A new entity among First Nations people of north western Ontario," *Canadian Family Physician* 42 (1996), 869–876.

11. See for instance, Maar et al., Sunday, Eyles, and Upshur, and L. C. Garro, "Intracultural variation in causal accounts of diabetes: a comparison of three Canadian Anishinaabe (Ojibway) communities," *Culture, Medicine and Psychiatry* 20 (1996), 381–420.

12. S. Folkman and J. T. Moskowitz, "Coping: Pitfalls and Promise," *Annual Review of Psychology* 55 (2004), 745–774.

13. Folkman and Moskowitz.

14. Aboriginal coping has been examined in several bodies of work by Iwasaki et al. See, for example, Y. Iwasaki, J. Bartlett, and J. O'Neil, "Coping with stress among Aboriginal women and men with diabetes in Winnipeg, Canada," *Social Science and Medicine* 60 (2005), 977–988.

15. The Maar et al. study suggested the patient's explanatory model of the causation of diabetes is one of several factors that influence Aboriginal people's coping behavior.

16. J. Frideres, *Aboriginal Peoples in Canada: Contemporary Conflicts* (Scarborough, Ontario: Prentice Hall, 1998).

17. M. E. Kelm, *Colonizing Bodies: Aboriginal Health and Healing in British Columbia 1900–50* (Vancouver: University of British Columbia Press, 1998).

18. N. Adelson, "The Embodiment of Inequity: Health Disparities in Aboriginal Canada," *Canadian Journal of Public Health* 96 s. 2 (2005), S46.

19. For a background on the UCCM see, http://www.uccm.ca.

20. Maar et al.

21. M. Maar, D. Manitowabi, D. Gzik, L. McGregor, and C. Corbiere, "Serious Complications for Patients, Care Providers and Policy Makers: Tackling the Structural Violence of First Nations People Living with Diabetes in Canada," *The International Indigenous Policy Journal* 2 no. 1 (2011), accessible at http://ir.lib.uwo.ca/iipj.

22. This approach to research is outlined in the "Guidelines for Ethical Aboriginal Research (GEAR)," Electronic document, http://www.noojmowin-teg.ca/default5.aspx?l=,1,613 (accessed 17 December 2009). For a description of the Seven Grandfather Teachings see E. Benton-Banai, *The Mishomis Book: The Voice of the Ojibway* (Hayward, WI: Indian Country Communications, Inc., 1988), 64.

23. Each First Nation varies in population. In 2002, the reported on-reserve populations were as follows: Sheguiandah 136, Whitefish River 265, Aundeck Omni Kaning 320, M'Chigeeng 872, Sheshegwaning 131, and Zhiibaahaasing 35. Source: Noojmowin Teg Health Centre, *Annual Report 2002/03* (2003), unpublished report by Noojmowin Teg Health Centre.

24. N. Adelson, *'Being Alive Well': Health and the Politics of Cree Well-Being* (Toronto: University of Toronto Press, 2004), 14–15. Adelson has also examined colonization as a form of social suffering, see N. Adelson, "Re-imagining Aboriginality: an

indigenous peoples' response to social suffering," *Journal of Transcultural Psychiatry* 37 no. 1 (2000), 11–34.

25. Adelson (2004), 114.

26. Ibid., 117.

27. The UCCM Health Board is named after this term, "Mnaamodzawin Health Services, Inc." Algonquian languages such as Ojibwa, Odawa, and Cree vary in dialect and classification. For instance, the linguist Richard Rhodes defines the Ojibwa-Odawa word for being "in good health" as *"mno-bmaadzid,"* see R. Rhodes, *Eastern Ojibwa-Chippewa-Ottawa Dictionary* (New York: Mouten de Gruyter, 1993), 494. On Manitoulin Island, *mnaamodzawin* translates as "good health." To contrast with the Cree term in Adelson's study (*"miupimaatisiiun"*) both phrases vary in dialect and classification (e.g., Odawa versus Cree), but culturally the meaning is the same.

28. There currently is no scholarly body of work that has examined this issue specific to Manitoulin Island. It is possible to infer this through such works as S. Pearen, *Exploring Manitoulin, Revised Edition* (Toronto: University of Toronto Press, 2006) and W. R. Wightman, *Forever in the Fringe: Six Studies in the Development of Manitoulin Island* (Toronto: University of Toronto Press, 1982). However the anthropologist Wayne Warry undertook a systematic study of health delivery on the north shore, north of Manitoulin Island, and linked colonial policies with community health. See W. Warry, *Unfinished Dreams: Community Healing and the Reality of Aboriginal Self-Government* (Toronto: University of Toronto Press, 1998).

29. Maar et al.

30. James V. Neel's "thrifty gene theory" posits that diabetes is genetically transmitted among Aboriginals and originates with the shift from feast and famine lifestyles to modernized lifestyles. See J. V. Neel, "Diabetes Mellitus: a 'thrifty' genotype rendered detrimental by progress?" *American Journal of Human Genetics*, 14 (1962), 353–362; and J. V. Neel, "The thrifty genotype revisited," in *The Genetics of Diabetes Mellitus*, ed. J. Kobberling and R. Tattersall (New York, Academic Press, 1982), 283–293. For a critique of Neel see M. Fee, "Racializing narratives: obesity, diabetes and the 'Aboriginal' thrifty genotype," *Social Science and Medicine* 62 (2006), 2988–2997; Ferreira and Lang, J. Poudrier, "The Geneticization of Aboriginal Diabetes and Obesity: Adding Another Scene to the Story of the Thrifty Gene," *The Canadian Review of Sociology and Anthropology* 44, no. 2 (2007), 237–261; and J. Poudrier, "Decolonizing genetic science: the thrifty gene theory, Aboriginal health and empowering knowledges," (PhD Thesis, Queen's University, 2004). For a critical medical anthropological perspective see H. Baer, "Steps toward an Integrative Medical Anthropology," *Medical Anthropology Quarterly, New Series* 4 no. 3 (1990), 344–348.

31. Poudrier (2004).

32. M. Hart, *Seeking Mino-Pimatisiwin: An Aboriginal Approach to Helping* (Halifax: Fernwood Publishing, 2002), 44.

33. Ibid., 48.

34. Ibid., 49.

35. Ibid., 50.

36. N. Scheper-Hughes, "Foreword" in Ferreira and Lang, xviii.

VI
____

PHYSICAL LANDSCAPES

11

Representing Indigenous Bodies in Epeli Hau'ofa and Syaman Rapongan

HSINYA HUANG

This chapter explores the representation of indigenous bodies within the context of the trans-indigenous Pacific. Drawing on the life narratives of Epeli Hau'ofa and Syaman Rapongan, two authors who are native to the Pacific region, I argue that Pacific indigenous body politics are very much connected to an "oceanic" body and constitute a "counter-conversion" from land to sea. In his thought-provoking book *Be Always Converting, Be Always Converted: An American Poetics*, Rob Wilson invokes the experience of "counter-conversion" as a reversal and turn away from Americanist codes of nation-state boundaries, a "developmentalist framework of global dependency and neocolonial entanglement in modern worlding" that stresses Native Pacific "smallness and belittlement" (121). All such hegemonic Americanism across the Pacific, colonial imposition of nationhood, race, boundary lines, and "false confinement into smallness, littleness, irrelevance, and global dependency" (127) could be turned over and cast away via "countervision of Pacific place-making and history-shaping" (130). If the United States is the hegemonic superpower that has converted the Pacific into "the American Lake,"[1] nowadays, as Wilson contends, Pacific Island authors and culture-makers figure the regenerations of "native attitude taking place across the contemporary Pacific otherwise and elsewhere" (120), which constitutes a "counter-conversion," "re-worlding" the Pacific through their "counter-visions and counter-memories" (119–120). To this, I would add that it is furthermore a "counter-conversion" from the continental to the insular, from land to sea, from "islands in a far sea" to an interconnected "sea of islands" alive with mobility, counter-mapping and counter-memory, and generative of action and community, as Hau'ofa figures it in his influential essay "Our Sea of Islands" (27–40). For Pacific island peoples, this entails belonging to the region through "sea-loving genes" and by "adoring the sea." In *We Are the Ocean* (Hau'ofa, 2008) as well as *Cold Sea, Deep Passion* (Rapongan, 1997) and *Black Wings* (Rapongan, 1999), Epeli Hau'ofa and Syaman Rapongan set out to liberate the indigenous body from the limitations of nationalistic discourse and social constraint; the respective resolutions of their life stories point us toward solutions to global problems of discrimination and ethnic or racial exclusion. In these works, the indigenous body is a cultural cipher, an index of (post)colonial history, as well as a figure upon which the cultural identities and life paths of Pacific indigenous peoples are inscribed.

Indeed, the body offers potential boundaries to the self; the works that I read here present the indigenous self as both a unique figure *and* a site for marking difference. Michel Foucault approaches the body as a product of discourse, which means that it virtually disappears as a material phenomenon. The physical, material body can never be grasped, for its existence is permanently deferred behind the grids of discursive meaning. In this chapter, however, the body is not only a basis for cultural identity and discourses of Oceania; it is substantiated and materialized in the experience of daily living. The body is the material and biological foundation upon which the superstructures of indigenous self and Oceanic community are founded. Through their own lived experience, as well as that of their island kin, Epeli Hau'ofa and Syaman Rapongan conceive of Oceania as a communal (sea) body, through which they can ultimately resist the imaginary political lines drawn by colonial powers. Their narratives turn hyper-modernized Pacific islanders (like themselves) back toward a perception of bodily identities as individual projects in intimate connection with Oceania.

Epeli Hau'ofa is one of the best-known contemporary Pacific island writers. His recent autobiographical work, *We Are the Ocean*, is a collection of essays, fiction, and poetry written over the past three decades. Like Hau'ofa's own frequent travels across the Pacific from his home in Tonga to Hawaii and Australia, the book's four parts, "Rethinking," "Reflecting," "Creating," and "Revisiting," represent a navigator's star chart of reflexivity and agency, criss-crossing Oceania. Landmark works such as "Our Sea of Islands," "The Ocean in Us," "Pasts to Remember," and "Our Place Within" outline his ideas for the emergence of a stronger and freer Oceania. Throughout the book, he expresses concern with the status of the Pacific islands and suggests that the most important role that the "people of the sea" can assume is as custodians of their ocean, the world's largest body of water.

Syaman Rapongan[2] is a Tao writer from Orchid Island, located forty kilometers southeast of Taiwan. There, tribal men and women continue to live a life close to their oceanic tradition, which is part of the larger Austronesian culture. He left Orchid Island for school when he was a teenager, remaining abroad for decades to work in the city as an urban indigene. There, he participated in the aboriginal demonstrations of the 1980s, of which the most significant was the Tao-led protest against the storage of nuclear waste on Orchid Island.

Like Hau'ofa, Syaman Rapongan returns in his writing to an oceanic perception of transnational community, islander space, myth, and language. The narratives of both authors revolve around the threat of being removed from their ancestral, natural surroundings. Colonial regimes have destroyed their lands with mining, deforestation, bombing, large-scale industrial and urban developments, and toxic dumping. This severs the indigenous population not only from their traditional sources of livelihood but also, and much more importantly, from their ancestry, history, identity, and ultimately their claim to a legitimate existence. As such, both authors advocate looking to their own cultural heritage to deal with the crises faced by their communities. Both advocate the perspective that their society and culture—indeed their very existence—are not fenced in by national boundaries. Rather, they and their people live in "[their] sea of islands," to use Hau'ofa's famous trope.

Syaman Rapongan's autobiographical writings also span the Pacific, intimately connecting the indigenous body with the oceanic sea body. The vibrant memory of the ocean is deeply rooted in the blood of his people, a memory that goes beyond genetic coding. With its innumerable waves of memory, the body is indeed a trope for cultural messages and historical meanings. The renowned Native American author N. Scott Momaday refers to memory as if it is stored in the body and passed down from generation to generation; he calls it the "memory in the blood." Coined in his Pulitzer Prize-winning novel, *House Made of Dawn* (1968) and developed in subsequent works such as *The Way to Rainy Mountain* (1969), "blood memory" is the metaphoric device that brings to life the stories of Momaday's grandmother:

> Although my grandmother lived out her long life in the shadow of Rainy Mountain, the immense landscape of the continental interior lay like *memory in her blood*. She could tell of the Crows, whom she had never seen, and of the Black Hills, where she had never been. I wanted to see in reality what she had seen more perfectly in the mind's eye, and traveled fifteen hundred miles to begin my pilgrimage. (*The Way to Rainy Mountain* 7; italics added)

In the tribal body is stored repressed indigenous memory, a racial memory that is embodied in a story in the blood. Such cultural coding exists beyond conscious remembering, so deeply engrained and psychologically imbedded as to be spoken of as "in the blood" (Weaver 8). Though critics such as Arnold Krupat remain wary of the essentialist dangers of blood logic,[3] I understand

Momaday's lexicon of the "memory in the blood" as an "evocative synonym for 'culture'."[4] What matters is not the biological blood, but rather indigenous cultural and communal milieu and ways of life. The genetic body serves as a metaphor for strengthened ties to cultural heritage. It voices a common tribal understanding of ancestral landscapes—in the case of Epeli Hau'ofa and Syaman Rapongan, seascapes and the universe as well—which provide access to a distinct culture and particular colonial experience, one that is incomprehensible to non-indigenes. The significance of body or blood memory resides in the heritage and identity passed down through story from generation to generation (Weaver 8).

The body is the memory that carries the law of equality and rebellion that organizes relations not only within an indigenous group, but between the group and their colonizers. It bears the colonial inscription and thus constitutes a material base for remembering a dislocated, displaced past. To regain a sense of identity in the face of enormous collective loss, the indigenous groups rely upon the memory, or story, that is preserved in their blood and body.

This *"tortured* body," as Michel de Certeau renders it, conflates with another body: *"the altered earth"*—a unity born of hardship and resistance to hardship inflicted by the colonial power (227). This communal body becomes the very site from which the will to *"construct* a *political* association" originates (227; italics original). It "deciphers the scars on the body proper [*le corps propre*]" as "the index of a *history* yet to be made": "Today, at the hour of our awakening, we must be our own historians."[5] Ideology is more often than not absent from indigenous demands; their actions are directed more toward the alliance of each community with a body and a land- or seascape. This allows real differences between their respective situations to be maintained, yet helps to establish the earth as common soil that can keep their "secret"—a "secret [that] remains uncompromised despite the alterations to which the Tablet of their collective law—the earth—is subjected" (229).

The idea of "memory in the blood" achieves "tropic power by blurring the distinctions between racial identity and narrative" (Allen 93–94). As Chadwick Allen argues, Momaday provocatively juxtaposes blood with memory and, in so doing, transforms the "taxonomy of delegitimation through genetic mixing into an authenticating genealogy of stories and story-telling" (94). He writes back to colonial discourse by re-inscribing an othered indigenous memory onto the body, thus embedding history into genetic codes. Blood memory, which has been forgotten yet never lost, redefines indigenous authenticity as recollection, and the genetic constitution preserves memory in the body. This is in effect an act of self-invention and empowerment, one that permeates indigenous texts as a functioning principle. Borrowing Momaday's phrase, I argue that indigenous peoples of the Pacific can claim the "Oceanic bloodlines" as a common inheritance. It bears colonial inscriptions, but it is also a mnemonic device. The memory in the body—both the individual and the sea body—keeps alive what European colonists tried to erase.

All this evokes the expansive, human-focused definitions of Oceania developed by Epeli Hau'ofa in a series of influential essays, published together in 2008 in the collection *We Are The Ocean*. In his seminal essay "Our Sea of Islands," originally published in 1993, Hau'ofa "offers a view of Oceania that is new and optimistic" by assuming the perspectives of "ordinary people" rather than those of "macroeconomics and macropolitics" (27). He concludes the essay with this stirring exhortation:

> Oceania is vast, Oceania is expanding, Oceania is hospitable and generous, Oceania is humanity rising from the depths of brine and regions of fire deeper still, Oceania is us. We are the sea, we are the ocean, we must wake up to this ancient truth and together use it to overturn all hegemonic views that aim ultimately to confine us again, physically and psychologically, in the tiny spaces that we have resisted accepting as our sole appointed places and from which we have recently liberated ourselves. We must not allow anyone to belittle us again, and take away our freedom. (39)

As opposed to the dominant view of "islands in a far sea," Hau'ofa defines Oceania as "a sea of islands with their inhabitants" (32). And indeed, there is a distinct difference between the two views. The first emphasizes dry surfaces in a vast ocean far from the centers of power, delineating their smallness, their remoteness. From this perspective, the islands are tiny, isolated dots. Around these dots, men from the continents—Europeans and Americans—drew imaginary lines, inscribing colonial boundaries that confined ocean peoples to tiny spaces. Today, these boundaries continue to delineate the island states and territories of the Pacific. One consequence of this view that the Pacific is a place where small and isolated groups of people occupy a vast surface, is that the region has been a favorite ground for weapons testing by all major powers (144).

People in the Marshall Islands, for example, have been victims of atomic and missile tests by the United States (31). The region has also been a site of toxic waste disposal and rapacious ocean resources exploitation. Nuclear-powered ships and vessels carrying radioactive materials ply the seas, while international business concerns seek islands for the disposal of industrial wastes. Activities that deplete the ozone layer continue in the region. Drift-netting, though it has abated, has not completely ceased, and the reefs of Moruroa Atoll may yet crack and release radioactive materials. People who are concerned with these threats try hard to enlist region-wide support to end them, but the level of their success to date has been modest (49–50). As a result, both indigenous and oceanic bodies bear witness to the parallel exploitation of the ocean and its people.

There has been, however, a shift from the protest stage to that of active protection of people and environment. Heading this agenda is a strong intention to redefine the indigenous experience and living place. The ancestors of Hau'ofa's "ocean peoples" have lived in the Pacific for over two thousand years. Their view of their world as "a sea of islands" is a holistic perspective in which things are seen in the totality of their relationships (31), where "Oceania" becomes a collective body. People raised in this environment are at home with the sea. They play in it as soon as they can walk, work in it, and fight upon its surface. They develop great skills for navigation, as well as the spirit to traverse the large gaps that separate island groups (32). The indigenous body merges with the sea body.

In Hau'ofa's narrative of "Oceania," the ocean is important not only for the stability of the global environment, but for the preservation of the indigenous human body. From waters that are relatively clear of pollution, Oceania meets a significant proportion of the world's protein requirement, produces marine resources, and provides global reserves of minerals. Such uses of the ocean are crucial to maintaining healthy human bodies and ensuring the sustainability of the environment. Together with their exclusive economic zones, the area of the earth's surface occupied by the Pacific island countries is vast. Kiribati, the Federated States of Micronesia, and French Polynesia, for example, are among the largest countries in the world by this reckoning. Indeed, there are indications that Oceania is becoming recognized for its pivotal role in the protection and sustainable development of the planet (37). We see this at the institutional level, with the establishment of organizations such as SPACHEE (South Pacific Action Committee for Human Environment and Ecology), SPREP (South Pacific Regional Environment Program), the Forum Fisheries Agency, and SOPAC (South Pacific Applied Geosciences Commission); in movements for a nuclear-free Pacific, the prevention of toxic waste disposal, and the ban on the wall-of-death fishing methods; and in the establishment of Marine Science and Ocean Resources Management programs at the University of the South Pacific, which have linkages to fisheries and ocean resources agencies throughout the Pacific and beyond.

Such indigenous agency in protecting the environment is evoked by Hau'ofa in his writing; he represents Oceanic peoples as custodians of the sea, who "reach out to similar people elsewhere in the common task of protecting the seas for the general welfare of all living things" (55). This may sound grandiose, but it is consonant with the growing importance of transnational movements to implementation of the most urgent projects on the global environmental agenda: protecting the ozone layer, forests, and oceans (55).

In his essay "The Ocean in Us," Hau'ofa refines the vibrant connection between the ocean and human bodies. For him, "Oceania" does not refer to an "official world of states and nationalities," but rather to "a world of people connected to each other" through Oceanic bloodlines (50). "This view," he argues, "opens up the possibility of expanding Oceania progressively to cover larger areas and more peoples" (51). Hau'ofa concludes this essay with the assertion that "the sea is our pathway to each other and to everyone else, the sea is our endless saga, the sea is our most powerful metaphor, the ocean is in us" (58).

It is important to acknowledge a theme of planetary connections in Hau'ofa's writing or, rather, to think of environmental belonging in planetary terms. His vision of Oceanic agency has its basis in what Lawrence Buell calls "eco-globalism," which involves a whole-earth way of "thinking and feeling" about environmentality (227). Buell proposes a planetary turn, which revolves around "ecoglobalist affect":

> By "ecoglobalist affect" I mean, in broadest terms, an emotion-laden preoccupation with a finite, near-at-hand physical environment defined, at least in part, by an imagined inextricable linage of some sort between that specific site and a context of planetary reach. . . . Ecoglobalist affect entails a widening of the customary aperture of vision as unsettling as it is epiphanic in a positive sense, and a perception of raised stakes as to the significance of whatever is transpiring locally in the here and now that tends to bring with it either a fatalistic sense of the inexorable or a daunting sense of responsibility as the price of prophetic vision. (232)

Specifying as it does the need to "reach out to similar people elsewhere in the common task of protecting the seas for the general welfare of all living things" (55), Hau'ofa's aspiration is in line with this "ecoglobalist affect," a planetary reach that intimately connects the near and far, inside and outside, human and non-human. Like Buell, Hau'ofa speaks of raised stakes when it comes to recognizing the global significance of local events; he transmits this through a poetic of relevance in which the far becomes relevant to the near, and vice versa, through a planetary interconnectedness. From this standpoint, Hau'ofa thinks against and beyond nation, conflating the individual body with the earth/sea body in his environmentalism. He is furthermore not only thinking, but *feeling*. The feel of the near-at-hand, and the sense of its connection to the remote is painful; this experience sets the ground for relevance and empathy: the effect of displaced and dispossessed indigenous groups, and of the devastation of their environment.

Epeli Hau'ofa came to see a past where people traveled unhindered. In this vision, the whole of Oceania was connected. Neighboring communities have always exchanged ideas and products, often across vast ocean distances, or between shore and mountainous interior. Both people and things traveled along these routes of interconnection, and it was a large world in which they mingled, unhindered by boundaries erected much later by imperial powers. From one island to another they sailed to trade and marry, expanding social networks and generating flows of wealth. In the past, Oceanic indigenes traveled to visit relatives in a variety of natural and cultural surroundings, "to quench their thirst for adventure, and even to fight and dominate" (33). Today, they do so again: everywhere they go—to Australia, New Zealand, Hawai'i, mainland United States, Canada, Europe, and beyond—they put down roots in new resource areas, secure employment and family property, and expand kinship networks in which relatives, material goods, and stories circulate across their ocean—and the ocean is theirs because it has always been their home (34).

Oceanic islanders have broken out of their confinement. They are moving around and away from their homelands, not so much because their countries are poor, but because they had been unnaturally confined and severed from many of their traditional sources of wealth. They move too because it is "in their blood to be mobile" (35). "Blood" is the key word. Citing Teresia Teaiwa, Hau'ofa proclaims, "we sweat and cry salt water, so we know that the ocean is really in our blood" (41). He continues:

> Oceania is us. We are the sea, we are the ocean, we must wake up to this ancient truth and together use it to overturn all hegemonic views that aim ultimately to confine us again, physically and psychologically, in the tiny spaces that we have resisted accepting as our sole appointed places and from which we have recently liberated ourselves. (39)

This expanded Oceania is a world of social networks that crisscross the ocean from Australia and New Zealand to the United States and Canada. By erasing the imaginary colonial mapping of the seascape, Hau'ofa constructs a discourse of the Oceanic (sea) body: "We are connected to Asia, to the Americas. I hope somehow in the future to make connections with America and places right around the Pacific, to tell our stories and see what we can do together" (39). In "our sea of islands," the immensity of a sea body without confines is configured as part of Oceanic genetic codes. His chapter titles ("the ocean in us," "pasts to remember," and "our place within,") use the body, and specifically blood, as tropes for his narratives.

What counts is not biological blood, but rather indigenous cultural milieu and ways of life that are deeply engrained in the body. The metaphor of the genetic body voices a common tribal understanding of ancestral landscapes, access to a distinct culture, and colonial experience. Landscapes and seascapes are cultural as well as physical, and Oceanic indigenes cannot read their histories without knowing how to read land- and seascapes alike. As expressed in Oceanic languages, the past is located ahead or in front, right there on the landscape, through a connection to past bloodlines. Speaking to a common experience of indigenous Oceanic peoples, Hau'ofa describes travelling through unfamiliar surroundings and being reminded of associations of particular spots or landscape features that were traversed in past events:

> We turn our heads this way and that, and right ahead in front of our eyes we see and hear the past being reproduced through running commentaries. And when we go through our own surroundings, as we do every day, familiar features of our landscapes keep reminding us that the past is alive. They often inspire in us a sense of reverence and awe, not to mention fear and revulsion. (73)

It is thus essential to protect the seascapes, for if they are destroyed, they take with them important memories, histories. It is significant in this regard that in several Austronesian languages, the words for "placenta" and "womb" are also the words for "land" and "sea." Hau'ofa's conflation of indigenous body with sea body through memories and histories is really closer to an "ethnoscape" of resistance (Appadurai 7),[6] and constitutes a counter-worlding to the colonial mapping of Oceania.[7]

Syaman Rapongan's memories are oceanic as well. It is with a similar oceanic sensibility that the Tao author writes:

> What does the "world atlas" mean? A chain of islands in Oceania. The islanders share common ideals, savoring a freedom on the sea. On their own sea and the sea of other neighboring islands, they are in quest of the unspoken and unspeakable passion toward the ocean or maybe in quest of the words passed down from their ancestors. (*Black Wings* 164)

The rich culture of the ocean implicitly deterritorializes the arbitrary and hegemonic boundaries of colonialism. The author depicts a huge Oceanic perception of place and space not only via an evocative sensibility for the ocean but also through a quest of the ancestral words passed down as bodily codes, as the memory in the blood. The ocean is reframed into an immense formation and Orchid Island is part of the interconnected islanders' heritage of Oceania. In fact, the center of the Tao world is not the island but rather the sea surrounding it. The vibrant connection between the ocean and the human is reflected in the related ceremonies of boat-building and boat-launching as well as the flying fish rituals, which Syaman Rapongan's narratives revolve around. The ocean is re-worlded through the poet's counter-vision and counter-memory as inherited from his ancestors. The indigenous subject acts to interpret and reconstruct Oceania as trans-border indigenous belonging. Though Syaman Rapongan left Orchid Island for a long and formative period, like Hau'ofa he returns to an oceanic perception of transnational island community. Both advocate the idea of living within their means, and cultivating body skills with the ocean. It is crucial for a Tao man to cultivate his bodily techniques: there are moments when he mysteriously retrieves memory through his body, memory that has been metaphorically composed in genetic codes and passed down from forebears in his blood:

> As I recall, for the last several years, I learned to dive and spearfish alone, trying to be a real Tao man to have the skills to support my family and to foster my self-confidence from the life experience of struggling with the ocean with the primitive physical strength of my ancestors. I tried to show my filial piety by providing fresh fish to my parents, to raise my children with sweet fish soup, just like my parents did when they raised me. (*Cold Sea, Deep Passion* 213)

In *Black Wings*, he describes this potent ancestral relation with the sea:

> My great-great-grandfather and all my forebears lived in this small island. The moment they were born, they fell in the love with the sea, entertaining themselves by watching, worshiping, and adoring the sea. The sea-loving genes are already contained in my body, passed down from generation to generation. I love the sea fervently, almost to the degree of mania. (80)

Syaman Rapongan's intimacy with that long-ago moment, with his beloved ancestors, is reminiscent of Hau'ofa's signature trope, "the ocean in us." For Syaman Rapongan as well, the links between oceanic and personal body empower the individual man, who can then become an agent of change. Syaman Rapongan formulates the capacity of the indigenous body to represent the hidden past and repressed memory, invoking the body as a site of vibrant connection and tribal knowledge. This ancestral intimacy as "the memory in the body," to again borrow Momaday's phrase, permeates Syaman Rapongan's narrative. It is sustained and strengthened by his cultivation of bodily proprieties and techniques, which allow him to interact with nature in a proper and traditional fashion. This bespeaks not only Syaman Rapongan's desire to "savor" the waves, but his urge to "save" the ocean.

Such bodily cultivation is made possible through the inheritance of bloodlines. When Syaman Rapongan and his father select a tree, cut it down and make it into a boat, the process is a sacred ritual and cultural lesson in traditional Tao knowledge of nature. Syaman Rapongan's father names the trees, delineating a rich tribal vocabulary as well as his knowledge of the various woods:

> Father (which is what "Syaman" means) of my grandchildren, this tree is apnorwa, and that one is isis. That tree is pangohen. . . . They are all excellent timber for building boats. This apnorwa has been waiting for you for more than a decade; it's the best timber for the middle pieced together on the two sides of the hull. This kind of timber rots the most slowly. This tree is a syayi, and it's the one we will cut down today for the keel. (*Cold Sea, Deep Passion* 56–57)

With prayers and poetry, Syaman Rapongan reveals the older Tao's attitude toward life, livelihood, and nature. His father mutters prayers as the son listens and learns:

> Oh, mountain god of the forest, I am a grandfather; you know my voice and the smell of my body. The father of my grandchildren also comes to pray to you with me. Don't let the knife and axe in our hands became dull, so that you can show off your hero-ism by breaking the surging waves on the sea. (*Cold Sea, Deep Passion* 57–58)

Syaman Rapongan's poetic sentiment revolves around humility and respect. By depicting his father and other tribal elders engaging in intimate conversation with mountain gods and tree spirits, he expresses an awe of nature and forges an itinerary for a bonding experience with it. Generational continuity is made possible here by the practice of responsibility and respect: "I tried to show my filial piety by providing fresh fish to my parents, to raise my children with sweet fish soup, just like my parents did when they raised me" (57). The acquisition of bodily skills for diving and spearfishing awakens Syaman Rapongan's repressed genetic memory. This is central to the transmission of Tao tradition, providing a platform for interaction between old and young, in which generations perform the practices of their culture together.

The rich ocean culture is the basis for Syaman Rapongan's deterritorialization of colonial boundaries: following the route of flying fish in *Black Wings*, Syaman Rapongan questions the very idea of "Taiwan-ness" and the nation-state's territorial sovereignty. In these passages, he configures lines of mobility and escape:

> The dense schools of flying fish dye patches of the wide and vast ocean black. Each school consists of three or four hundred fish,

swimming about fifty or sixty meters apart. They stretch unbroken for one nautical mile and they look like a mighty military force going into battle. They follow the ancient course of the Black Current, gradually heading toward the sea north of Batan in the Philippines. (5)

I am like a stone that is about to burst; I cannot help but sing out the *Mivaci* (Song of the Catch). As I start to sing, I cannot stop. I sing from my heart. The open sea of Hong Kong and Macao vibrates with the song of the *Mivaci*. Oh, the singing far and near, as if oppressed for a thousand years, spurts out from the black sky and the bottom of the black ocean, to awaken all the people of the tribe. (39)

We can only see Mina-Mahabteng (Pisces) in winter—Mahabteng, you know, is the kind of fish you catch in the oceanic trench and reef coastline. Long ago, the constellations of Mina-Mahabteng directed our ancestors back to this island from Batan in the Philippines, where they did business with its residents. The constellations are located right above the mountains to our north. (183)

The "black wings" of the flying fish return every year, inspiring the island-ers' will to survive and serving as fountainhead of their fighting spirit. The fish—indeed, the very waves—carry memories of Tao ancestors. The inter-island migration of these ancestors formed a "partial community," to use Homi Bhabha's phrase, rendered by locally mediated kinship. In Wai Chee Dimock's words, Homi Bhabha's "partial community" of the global/planetary life is exactly "rendered partial by its off-center relation to the national govern-ment, and by its far-reaching and locally mediated kinship with other distant minority groups" (11). In such circumstances, the minority, always a partially denationalized political subject, emerges as a "partial and incipient" (Bhabha, "Statement for the *Critical Inquiry* Board Symposium" 342) social force that seeks to recognize itself and represent its freedom through identification with the difference of the other—its claims, interests, and conditions of life. Such identification converts the liminal condition of the minority—again, always partly denationalized—into a new kind of strength based on the solidarity of the "partial" collectivity, rather than sovereign mastery. "This is a subset of humanity that cannot be integrated into a sovereign whole, a subset always partly external to any nation-based set," as Dimock puts it (11).[8]

Like "the dense schools of flying fish," the diverse island inhabitants cross over and pass over the ocean, following the natural rhythm of the Black Current. They recognize no borders. This act of border-crossing characterizes

Syaman Rapongan's tribal indigenes as it does the inhabitants of other Pacific islands. As Hau'ofa makes explicit in *We Are the Ocean*,

> [The islanders] cannot afford to believe that [their] histories began only with imperialism. . . . [They] cannot afford to have no reference points in [their] ancient pasts to have as memories or histories only those imposed on [them] by [their] colonizers and the present international system that seems bent on globalizing [them] completely by eradication [of their] cultural memory and diversity, [their] sense of community, [their] commitment to [their] ancestry and progeny, and individualizing, standardizing [them], and homogenizing [their] lives. (76)

Syaman Rapongan, like Hau'ofa, subverts hegemonic borders by following the movement of the Black Current, which shapes and reshapes the migratory route of his ancestors as well as that of flying fish. "[I]n a constant state of transition," to use Anzaldúa's phrase (25), those who move through imperial confines reset borders as lines of dispersal, escape, and resistance. As they trespass they also transgress, challenging the rigidity and normality of border constraints. Bound together by history, memory, land, and spirit, Pacific island indigenes exert their power of resistance, deterritorializing national borders into lines of mobility and escape.

Both Syaman Rapongan and Epeli Hau'ofa challenge the adequacy of a nation-based paradigm, using oceanic vocabulary and insular metaphor. Both track the colonial degradation of nature and mapping of vast seascapes, translating ecological crisis into personal and collective life narratives. Their common theme is the interconnectedness of the individual indigenous body to an un-demarcated sea body. This connection expands from within to beyond, from Oceania to "Asia, to the Americas, from Australia and New Zealand in the southwest to the United States and Canada in the northeast" (Hau'ofa 39). Likewise, Syaman Rapongan follows the black current of the ocean to move beyond his isolated island, and in doing so makes a connection with his ancestral Austronesian tradition. Both authors set out to represent a voice from the margin, and end up configuring an indigenous geography—a sea body of interconnectedness that bypasses the boundaries of the nation-state and re-envisions the trans-indigenous as a relationship of center to center.

Pacific tropological energy is impressive. Islands in a far sea are reframed into an interconnected "sea of islands," alive with mobility and metaphor, generating action, community, and hope. "Oceania," refigured from the interior Pacific, becomes a way of building up a new form of Pacific Islander "oceanic consciousness," an almost Freudian concept of primordial bodily unity between the self and other. But this time, the consciousness is a consolidation from below, framed in a physical-geo-politicized mode. The "cross" gives way to the "crisscross" of trans-Pacific mobility and interconnection, reciprocity of indigenous body and oceanic sea body. As Rob Wilson insightfully puts it, Oceania becomes not just heritage or blood, but "a trope of commitment, vision, and will" (15).

Notes

1. Rob Wilson appropriates the image of the "American Lake" from John Pul's *The Shark That Ate the Sun* to address the eco-political catastrophes faced by the Native Pacific Islands (119–120). As he affirms, John Pul "protests with voices rising from the nuclear graves, 'the American Lake'" (120). Pul's poem is cited as an epigraph to the chapter "Writing down the Lava Road from Damascus to Kona":

> I look at the map of the Pacific.
>   The American navy calls the Pacific the American Lake.
> They have ships in Samoa
> Hawai'i, Taiwan, the Philippines,
> Belau, Kwajalein, Truk
> The Mariannas, the Carolines,
> In Micronesia, there are 90,000 people,
> Who gives a damn?
> The dead are louder in protest than the living,
> The living are silent.
> Silent.
> Silent.
> Silent.
> Silent.
> Silent.

2. "Syaman" in the Tao language means "father." "Rapongan" is the given name of Syaman's eldest son; therefore, "Syaman Rapongan" literally means "Rapongan's father." The naming customs of Tao people and other Taiwanese indigenes are distinct from the majority Han culture. The Tao consider giving birth to be a major achievement in life. In order to commemorate the birth of a new generation, they inscribe this passage of life in their names. If the name starts with "Si," that means this Tao person is still single. If someone becomes a father, he changes his name to "Syaman," followed by the given name of his eldest son or daughter. If someone becomes a grandfather, he changes his name to "Syaben," followed by the given name of his eldest grandson or granddaughter. For details of the naming systems of indigenous people in Taiwan, refer to Mayaw Biho's documentary series "What's Your Name, Please?"

3. In *Red Matters*, Arnold Krupat derides Momaday for his use of the phrases "racial memory" and "memory in the blood" (76–97). He attacks the essentialist

dangers of references to "Indian blood," asserting that Native American resistance should not be associated with the vexed category of "Indian blood," which is roughly parallel to the concept of race (xi). Krupat argues that to dismantle the intricate edifice of racism, it is imperative to expose the essentialism embodied in the idea of "Indian blood" and discard its associated policies. "Indian blood" in effect appears to Krupat as "a discourse of conquest," as he cites Pauline Turner Strong and Barrik Van Winkle ("'Indian Blood': Reflections on the Reckoning and Refiguring of Native North American Identity," *Cultural Anthropology*, 11 no. 4 (1996): 565; qtd. in Krupat 76). Even in the interest of rights and resistance, Krupat remains wary of blood logic.

4. See H. David Brumble, III, *American Indian Autobiography* (Berkeley: U of California P, 1988); qtd. in Weaver 7.

5. DIAL Document no. 196; qtd. in de Certeau 227. This is part of the address by Justino Quispe Balboa before the first Indian Congress of South America, October 13, 1974, in the presence of Paraguayan authorities.

6. Arjun Appadurai understands the ethnoscapes as "the landscapes of group identity," the globalized spatial diffusions of ethnic communities, which are no longer bound to certain territorial locations. Appadurai creates the neologism "ethnoscape" to describe a transnational and intercultural phenomenon, deriving from global changes in human society. The term is in particular useful in investigating the linkage between space and ethnic perceptions. For details, refer to Appadurai's chapter "Disjuncture and Difference in the Global Cultural Economy" in his renowned study on globalization and diaspora, *Modernity at Large: Cultural Dimensions of Globalization.* Minneapolis: U of Minnesota Press, 1996.

7. "Worlding" is a concept Gayatri Charkravorty Spivak develops quite extensively in her writings. In her landmark essay, "Three Women's Texts and a Critique of Imperialism," Spivak delineates the "epistemic violence" in the novel *Jane Eyre* (and in the case of Bertha, physical containment and pathologization). Spivak portrays such imperialism as a "worlding" process that attempts to disguise its own workings so as to naturalize and legitimate Western dominance:

> To consider the Third World as distant cultures, exploited but with rich intact literary heritages waiting to be recovered, interpreted, and curricularized in English translation fosters the emergence of "the Third World" as a signifier that allows us to forget that "worlding," even as it expands the empire of the literary discipline. (269)

Hau'ofa's writing constitutes a "counter-worlding" in the sense that he evokes strong local sensibility for Oceania and translates Oceania as a figure, spacious and interconnected, which counteracts colonial imposition and mapping.

8. Using the language of set and subset as a heuristic guide, Dimock designates the field of American studies as a "continuum rather than a container," suggesting "a porous network, with no tangible edges, its circumference being continually negotiated, its criss-crossing pathways continually modified by local input, local inflections" (3). It allows us "to modularize the world into smaller entities able to stand provisionally and do analytic work, but not self-contained, not fully sovereign, resting continually and nontrivially on a platform more robust and more extensive" (4).

Works Cited

Allen, Chadwick. *Blood Narrative: Indigenous Identity in American Indian and Maori Literary and Activist Texts*. Durham: Duke UP, 2004.

Anzaldua, Gloria. *Borderlands/La Frontera, the New Mestiza: Third Edition*. San Francisco: Aunt Lute Books, 2007.

Bhabha, Homi. "Global Minoritarian Culture." In Wee Chi Dimock and Lawrence Buell (eds.) *Shades of the Planet: American Literature as World Literature,* pp 184–95. Princeton: Princeton UP, 2007.

————. "Statement for the *Critical Inquiry* Board Symposium," *Critical Inquiry* 30 no. 2 (2004): 342.

Biho, Mayaw. "What's Your Name, Please?" Documentary Series. Taipei: Public Television Service Foundation, 2002.

Buell, Lawrence. "Ecoglobalist Affects: The Emergence of U.S. Environmental Imagination on a Planetary Scale." In *Shades of the Planet: American Literature as World Literature*. Ed. by Wee Chi Dimock and Lawrence Buell, 227–48. Princeton: Princeton UP, 2007.

de Certeau, Michel. "Politics of Silence: The Long March of the Indians." In *Heterologies: Discourse on the Other*, 225–33. Minneapolis: U of Minnesota P, 1986.

Dimock, Wee Chi. "Introduction: Planet and America, Set and Subset." In *Shades of the Planet: American Literature as World Literature*, edited by Wee Chi Dimock and Lawrence Buell, 1–16. Princeton: Princeton UP, 2007

Foucault, Michel. *Language, Counter Memory, Practice*. Ithaca: Cornell UP, 1980.

Hau'ofa, Epeli. *We Are the Ocean: Selected Works*. Honolulu: U of Hawaii P, 2008.

Krupat, Arnold. *Red Matters: Native American Studies*. Philadelphia: U of Pennsylvania P, 2002.

Momaday, N. Scott. *The Way to Rainy Mountain*. 1969. Albuquerque: U of New Mexico P, 2001.

Rapongan, Syaman. *Cold Sea, Deep Passion*. Taipei: Lianhe Wenxue, 1997.

————. *Mythology of the Bay of Eight Generations*. Taichung: Chenxing, 1999.

————. *Black Wings*. Taichung: Chenxing, 1999.

————. *The Memory of the Waves*. Taipei: Lianhe Wenxue, 2003.

Spivak, Gayatri Chakravorty. "Three Women's Texts and a Critique of Imperialism." In *"Race," Writing, and Difference*. Edited by Henry Louis Gates, Jr., 262–80. Chicago: U of Chicago P, 1986.

Teaiwa, Teresia. "Reading Paul Gauguin's *Noa Noa* with Epeli Hau'ofa's *Kisses in the Nederends*: Militourism, Feminism, and the 'Polynesian' Body.'" In *Inside Out: Literature, Cultural Politics, and Identity in the New Pacific*. Ed. by Vilsoni Hereniko and Rob Wilson, 249–64. Maryland: Rowman & Littlefield Publishers, 1999.

Weaver, Jace. *Other Words: American Indian Literature, Law, and Culture*. Norman: U of Oklahoma P, 2001.

Wilson, Rob. *Be Always Converting, Be Always Converted: An American Poetics*. Cambridge, Mass.: Harvard UP, 2009.

12

The Many Indigenous Bodies of *Kai Tahu*

KHYLA RUSSELL AND SAMUEL MANN

What is the substance of this Māori cultural identification with
landscape and coast, with water and mountain, with species and
resources? At the core of the Māori view of landscape is whakapapa,
[that] which connects people to the[ir bodies of sea and] land.[1]

This chapter will explain and illuminate the terms that we as *Kai Tahu*[2] use
to describe and locate ourselves and our bodies in our world. In a Treaty
relationship with Otago Polytechnic, designed to further teaching and *ako*
(learning) and *rakahau* (research), *Kai Tahu* has participated in the SimPā:
Digital Landscapes of Māori Memory project.[3] This is an IT program that has
the potential to recreate in virtual reality, land and seascapes using *Kai Tahu*
ancestral stories. The SimPā project enables *Kai Tahu* participation in digital
media design, using 3D game technology to encourage and promote distinctly
Kai Tahu voices, stories, and cultural content. This project brings together
two key components: stories and technology. Most importantly, the project
facilitates intergenerational Māori interaction through new media, reinvigorates interest in educational processes, and provides professional training for
a new generation of Māori digital specialists.

As an *iwi* (tribe), we have connections through our bodies with our
hinekaro (psyche or mind maps), with our *wairua* (the unseen or spiritual
aspect of our bodies), with our *whānau* (immediate and extended family),
with our *hapū* (sub-tribes), and with the *iwi* (larger tribal community): those
living, yet to be born, and dead. This is how we know who we are, how we
connect ourselves (through *whakapapa*[4]) to people and landscapes and seascapes, which are cosmological, physical, and cultural in conceptual terms.
It is also as people who conceive of these places as being part of themselves,

and from which we derive our bodies of knowledge, both in terms of specific knowledge domains (*kauae a ruka* or "upper jaw knowledge") and everyday knowledges associated with survival, food production, and caring for people and resources (*kauae a raro* or "lower jaw knowledge").

We connect physically with our land- and seascapes by living on them, and by humanizing them when we do not live on them directly. Over time, people's perceptions of these land- and seascapes have been altered by human and natural actions upon them. Our indigenous literature in times past was, and in many ways continues to be, expressed through *whakairo* (carving) and painted *kowhaiwhai* (specific *Iwi* patterns based on nature often presented in the abstract); through *tā moko* (bodily tattoos[5]), by way of *tātai kōrero* (stories[6]), and lastly by use of and engaging in *kōrero neherā* (ancestral stories[7]). Thus landscape stories have been retained through endless generations as bodies of knowledge lovingly held as precious by keepers charged with their *kaitiaki* (guardianship).

Kai Tahu educational aspirations include the ability to aim for lofty heights. Through our ancestral knowledge, we are inspired to rise to such challenges when we think positively, so we can soar with our dreams like one of our great birds, the toroa (Royal Albatross). This bird is spoken of in another of our *whakatauki* (sayings), which tells how the *Kai Tahu* messenger birds ask us "who are these messengers?"[8] Application of these two related *whakatauki* underpin the basis for our partnership with Otago Polytechnic, as do two forms of Intellectual Property (*Iwi* and academic), so we aim to have only positive outcomes for all participants. In the SimPā project, we use established and newer media to reiterate and reaffirm our particular form of indigenous storytelling of pre-histories and histories of our places, as part of the wider bodies of our *Iwi* landscapes. Through this project, *Kai Tahu* aspire to carry messages of significance to our *Iwi* members. Using contemporary technologies we can retain our SimPā created stories using new means of imparting them to our *Iwi*, wherever in the world they may be located. The project is also about safeguarding and protecting our knowledge and knowledge systems by using new means of dissemination from that of the messenger birds discussed above. Our Treaty of Waitangi (1840) promised to forever protect absolutely our treasures and all things we hold precious, and the stories that have been re-created are placed squarely within *Iwi* epistemologies, ontological understandings, and *mātauraka* (knowledge systems). This has engaged us with information technologies and confirmed the value of our Treaty partnership with Otago Polytechnic.

Treaty partnership operational models require specific *Iwi* aspects and considerations. In this instance, the three Cs of connectedness, co-operation (or collaboration) and complementarity underpin our model. Using these as the model, this chapter will demonstrate how *Kai Tahu* and Otago Polytechnic have entered into one such treaty relationship to engage in teaching and *ako* (learning) and *rakahau* (research). Both partners have benefitted in new and innovative ways.[9] This project and partnership arose because one of the areas in which *Iwi* Māori generally underachieve is in compulsory and tertiary education. This chapter will not discuss the root cause of these failings, but look instead at a collaboration that is having positive outcomes, and thus countering numerous commissioned pieces of research that continue to tell us of the failures while none address the causes.

Initial discussions between Otago Professors Samuel Mann and Khyla Russell, who is also *Kai Tahu*, suggested a way to involve both young and older *Kai Tahu* in education and learning about themselves and their cultures. A new and engaging way of interesting *Kai Tahu* youth was through the concept of SimPā. The end product of the discussion was a resource that the partners (*Iwi* and Polytechnic) could co-develop. For *Kai Tahu*, this meant new ways of imparting knowledge and new ways of recreating it in virtual reality. This project also aimed to create new ways to foster and strengthen research between Māori and academic communities in mutually respectful ways. Two years of discussion between *Kai Tahu* and Otago Polytechnic about terms of reference and the requirements of external Acts of Parliament produced a Memorandum of Understanding. This ensured that any collaboration in research would prove educationally and culturally meaningful for *Kai Tahu*, and protect intellectual property rights for both partners. The result brought a moment of true potential, from which emerged the SimPā technology.

Aims and Objectives of the SimPā Project (2006)

The SimPā project aims were to convey and strengthen *Iwi* culture and knowledge by initiating a process of participatory Māori digital media design using 3D game technology. This project recognizes that Māori culture is a vital part of what distinguishes New Zealand from the rest of the world. The project aimed to enable the creation of 3D game-based Māori digital content, so that distinctly Māori voices, stories, and cultural content can be encouraged and disseminated. This development, it was hoped, would have technological and cultural benefits, and that the two could be fused in *Iwi* digital content.

The project aimed to achieve its goals through *Kai Tahu* active engagement and participation. Its purpose was to develop a process of participatory game development for Māori cultural content; to develop a SimPā[10] "toolkit" to enable this; to develop structures for use of resultant GamePā, a virtual environment representing a specific place; and to develop a new capability for training digital storytellers. There were three major justifications for this project. First, the risk of *Kai Tahu*-specific Māori knowledge being lost due to the reduction of knowledge repositories. Second, the impact of negative statistics of educational outcomes for Māori within compulsory education, and the lack of engagement of young *Kai Tahu*. Third, and as a direct consequence, the lack of skilled practitioners with the ability to input *Iwi* Māori digital content.

The dual need for content and capability is recognized by *Iwi* as both a limitation and an opportunity, as "communication technology is providing new avenues for our people to be enriched and [to] contribute to our *Kai Tahutaka* regardless of time and location."[11] The importance of Māori digital content is also key to the government's Digital Strategy, as *Iwi* Māori culture is a vital part of what distinguishes New Zealand from the rest of the world. Information and Communications Technology (ICT) can be used to help create the conditions for the realization of the diverse forms of *Iwi* Māori potential. It is crucial for the future of *Iwi* Māori and of New Zealand as a whole that distinctively Māori voices are encouraged and promoted. It is important for New Zealanders from all walks of life to be able to create and use their own digital content in order to create social, cultural, and economic value for themselves, their communities, and their nation.

Iwi are both creators and consumers of content, and distinctively Māori content is particularly visible in the areas of broadcasting, the arts and creative industries, education, health, and business sectors, including tourism. Māori digital content is important not simply for its economic potential, but also as a vital means of expressing Māori culture in today's society and into the future, strengthening Māori society and identity, telling Māori stories to other Māori, and communicating with the wider world. Hence the importance of content being created and maintained in the Māori language.

As part of the funder requirements, and in order to have a Māori-focused outcome for *Rūnaka* (local tribal councils) rather than simply research outputs, the following objectives were necessary. First, the development and testing of a participatory approach to game development. This brings together people from within the *Papatipu Rūnaka* (localized tribal council), who will jointly learn about their own place and stories, and convert this knowledge to digital form. Second, the development of a software tool for creating specific Māori virtual environments, i.e., the "SimPā toolkit" (*he kohinga o ngā mea rauemi*). Third, the development and testing of tools for the use of games in teaching Māori concepts. This encompasses specific research on the effectiveness of digital game based learning in a Māori context, i.e., the use of the GamePā. Fourth, the development of techniques and practices for the further use of GamePā. The resulting games will provide an interactive learning environment for use within each *Papatipu Rūnaka* (localized tribal council). It is expected that this will enhance their *mātauraka* (knowledge systems) and enable individuals to connect with, and have knowledge of, their landscape and historical stories. Each game will be the intellectual property of each *Papatipu Rūnaka* and provide an indigenous tool for future development in education and Māori business. We believe that this integrated model—using a resource that is interactive, online, and multiplayer—will provide measurable benefits for individuals, for our *whānau* (immediate and extended family), for our *hapū* (sub-tribes), and for the *iwi* (larger tribal community). Fifth, the development of a new specialist area in education: Māori digital content. We are developing a program aimed at capacity building among indigenous people. By combining cultural knowledge with skills required for developing digital game based learning resources, we hope to initiate a pathway to encourage careers in this area. This will provide career opportunities in education, information technology, and business. Sixth, the development of this initiative beyond the collaborating partners. This collaboration is not just between a single group of stakeholders, but involves complex structures of knowledge ownership. An important part of this initiative is the development of processes maintaining the integrity of specialist knowledge and *tikaka* (appropriate cultural and interpersonal behavioral[12]). Specific *Kai Tahu* concepts were always refined for presenting back to Otago Polytechnic.

Due to generations of urbanization, different uses of land and altered landscapes, the Māori knowledge with which we engaged—*kōrero neherā* (ancient histories and specific knowledge)—was either lost, or in danger of being lost. As with countless thousands of indigenous groups worldwide, the landscapes of our *tūpuna* (ancestors) were lost to new ownership. Samuel Mann's idea was to recreate virtual landscapes and their histories using information technology and marrying that with our *kōrero neherā* (oral recited ancient histories[13]). Thus began a serious consideration by both Otago Polytechnic and *Arai-te-Uru Rūnaka* (local tribal councils). The idea needed discussions with *Iwi* members who knew the oral histories, who might be interested in collaboration, and who had time, commitment, and energy to help drive this. Software developers and *Iwi* experts worked together in the first instance to create a model to be offered to *Iwi*. This first tester model visualized a *hui* (tribal building/living complex[14]) and included a series of specific learning workshops on *wānaka* (*Iwi*/Māori-based knowledge[15]), and considered offering—through a series of *Iwi*/Māori knowledge focused weekend workshops—a qualification when the resource was completed. Despite the absolute success of the tester, the group whom we thought or had hoped would show an interest were not those who had attended.[16] In fact, this event interested people much like us: people who saw its potential as a teaching and learning tool and, in some cases, who had travelled huge distances to attend. With Samuel's enthusiasm, and with the successful funding application which helped to make it a reality, SimPā became a research collaboration.

Early on, Samuel, his programmer, and the potential *Kai Tahu* lecturer visited with tribal councils, attended their meetings, presented the tester to them, and did the same at the *Iwi* head office. Each time, the project was received with enthusiasm for the potential it offered *Iwi* Māori, regardless of age. The potential teaching and learning was intergenerational which fitted within *tikaka–a-Iwi* (traditional *Iwi* understandings of appropriate cultural and interpersonal behavioral). Grant monies enabled us to make a 0.8 research project appointment. The person appointed was *Kai Tahu*, and his/her role was to ensure the making/gathering of resources in collaboration with *whānau* (family), with *hapū* (sub-tribes), and with those interested in sharing their knowledge.

Intellectual Property

A crucial issue within the project has been the recognition of the importance of different forms of knowledge. In this context, the computer program is owned by the university; the games it contains are owned and controlled by the local tribal councils.[17] The exception is in the evolutionary development of the SimPā toolkit. The toolkit is a key component of the project, a software tool for creating virtual environments specific to Māori culture and understandings of appropriate behavior. The toolkit is a continually expanding collection of models, artwork, images, sound files, and programming scripts that can be assembled together to define the appearance of a virtual environment, and how a user can experience and interact with it. The toolkit simplifies and accelerates the process of creating a GamePā: a virtual environment representing a specific place. Each tribal council involved will produce GamePā.

Developing a framework for representing indigenous knowledge through a games-based environment is useful not only for Māori, but also has international appeal for indigenous groups in Canada, the United States, and Australia, etc. Potentially, there is significant commercial value in this model, and we believe that a positive outcome of the project is that it can be piloted and perfected locally for the benefit of the community, then commercialized in partnership with the initial stakeholders.

Management and Governance

This has been a project of interactions and overlapping responsibilities, with various committees involved from, and on behalf of, both partners. The project reported to a range of other boards, and reporting fell into four main areas: governance; the evaluation of *tikaka* (appropriate cultural behavior);

the evaluation of ICT components; and financial management. With such safeguards in place, we made great inroads even though it was often difficult to negotiate agreement between tribal groups on stories and who could be storytellers.

One tribal council has been involved in telling migration histories, and has held a series of learning workshops over a number of months. The stories have been filmed, landscapes recreated, and family members and workshop participants have had the potential to learn the additional skills of video and film making. Through simulating landscapes to adorn their histories, workshop participants partake as learners in cultural activities which teach new histories of the places they know from a tribal historical perspective.

Between these workshops, editing and further meetings happen to ensure that appropriate processes are engaged in and transparency of behavior, knowledge transition, and intellectual properties are protected and acknowledged. One family group is involved in replanting and re-establishing a wetland on their *wakawaka* (family land). Some of our SimPā staff are attending these workshops which include cultural, traditional, and botanical education; *kai Māori* (traditional food), *karakia* (incantations[18]), and behaviors appropriate to planting, knowledge, and food gathering. Another family group in adjacent family land is involved in recording their oral histories, their *mahika kai* (food gathering and production) knowledge, their *whakapapa* (genealogical ties) and re-creating aspects of all of these through *whakairo* (carvings). The second carving is a memorial work begun by one of their members to honor his father and a younger brother who drowned while fishing.[19] Accompanying *karakia* (incantations), *kōrero* (narratives), *kaupapa* (ideas), and the *whakaaro* (thinking) informing what is *tika* (correct), are incorporated into the learning by all partners in the collaboration. As each set of these resources was made, it was edited and produced in an agreed format. Those whose stories these are will decide how they may be distributed and to whom. Some have appeared on the social networking Web site BEBO, some on YouTube and all have been made in DVD formats. Additionally, tribal councils have had tutorials on their use and how to add to the resources.

Once workshops have occurred, there is always a huge exchange of knowledge, information gathering, and a sharing of expertise. This expertise includes traditional Māori knowledge and using IT systems to best complement the old knowledge and make it available in the present. We now have what once might (and often still is) be seen as art history, anthropology, prehistory, navigation, food production and gathering, lunar and environmental knowledge, and teaching and learning as the basis of our teaching and *ako* (learning). All of these knowledge forms have been located within information technology within a research project which can help inform educationalists nationally and internationally about innovative and alternative engagement in teaching and learning that can engage young and older community members alike. The project required ongoing collaboration, to achieve cooperation, connectedness and complementarity between both partners, *Iwi* and university. A number of other tertiary and related institutions, both *Iwi* and nationally, plus *Iwi* at home and abroad, have expressed an interest in how they may be able to access this knowledge. Most are keen to know about the processes in which we engaged prior to the project beginning.

This was never more evident than in October 2007, when Samuel Mann and Khyla Russell presented a seminar on our progress to date at London University on our way to a conference in Toronto. A number of *Iwi* Māori and others from New Zealand were keen to engage in order to see what we might, in the future, offer them. *Iwi*, especially, were wanting to access the information about how we had negotiated with *Kai Tahu* tribal councils so that they, in turn, might be able to take their homeplaces and *Iwi* stories to their children living in London. The cost of travel for whole families is so prohibitive that a number feared they and their children would become *manene* (strangers) in their ancestral places.

In both London and Toronto, different audiences were keen for ongoing or further contact, for more presentations, and for more information on how the agreement between nations might come to fruition from a research perspective, or as a means of engagement with their first nations people.[20] Khyla was able to be part of the introductory evening of poetry, performance, and *haka* (dance), as well as food. This introductory evening was a prelude to the fiftieth anniversary celebrations of the founding of the London Māori Club, the *Ngāti Ranana*. Part of Khyla's role was to be a conduit for receiving and giving news of home and people; for sharing food and *kōrero* (narratives); for being with our *whānau* (immediate and extended family) and *hapū* (sub-tribe), and for bringing home these *karere* (messages), including SimPā. In time, we aspire to fulfill many of these aspects through SimPā, so that more regular cultural and *Iwi* engagements are possible. Technology from far-off places has always been accepted and integrated, and began with whalers and traders before settlers arrived in New Zealand. Inter/exchanges are what *Iwi* still desire, however these opportunities may present themselves.

In this example, information technology is the essential tool that sits alongside programming expertise, and indigenous knowledge to create and then broadcast this information. For *Kai Tahu*, it has provided another means of being able to work with Otago Polytechnic and its rich technical and human resources, in order to produce accessibility to learning entry for our people. We now see computer scientists, computer programmers, their students, and our *Iwi* members at family, sub-tribe, and tribal council levels, engaged in educational pursuits in some or all of the above ways. Film makers/editors and video operators teach our tribal members to use these tools, and we in turn teach *Kai Tahu* our landscape histories as told by our older members to our younger ones. We now have a means to access, perpetuate, and disseminate our *kōrero neherā* (ancestral histories) in new and innovative ways.

These interactions are always precipitated by our long-practiced rituals of encounter and add deeper dimensions to our histories and pre-histories. Knowing the ways we continue to engage formally in spaces (*marae atea*) designated for rituals of encounter,[21] attending *hui* (tribal gatherings), and accessing traditional as well as contemporary learning, means we can begin to experiment and more accurately guess how ancient landscapes might have looked using *waiata* (ancient chants) and their accompanying stories. In this way, culture can be experienced in a virtual space more meaningfully than by merely reading an anthropological or historical account. This virtual Māori way means it is grounded in *tikaka* (appropriate behavior) and *kawa* (fixed cultural protocols[22]).

The programming experts help the participants to understand how to create digital images, games, and DVD presentations of the virtual realities of people and places of our *tūpuna* (ancestors), acting and interacting in these virtually created spaces. Our people far from home can access these histories and *te reo* (language) of their places, ancestors and their deeds; and give this new knowledge to their children, many of whom have never set foot on these landscapes. Applying the three Cs—connectedness, co-operation (or collaboration) and complementarity—has given us access to two cultures that have found balances, made choices and made positive connections between traditional and mainstream education. Each partner has begun to become more expert in how to cooperate and collaborate during this process.

Both Sam and Khyla, in different educational arenas and various conferences and similar fora, continue to learn about how other institutions locally, nationally and internationally, engage or fail to appropriately engage with their indigenous peoples. Through collaboration, we have continued to experience new ways of coming together to make new resources, using new means of receiving education, and giving or offering it to our own *Iwi* who live in far off places. True and accurate *Kai Tahu* geo-cultural images (*ētahi whakaahua ki a Kai Tahu ki Ara-i-Te-Uru*)[23] use SimPā as a tool to make access possible via the Internet, through the creativity of using and accepting virtual reality as a means of connecting.[24]

Notes

1. T. O'Regan, unpublished *Runaka* Calendar, *Te Runanga O Ngai Tahu*, 1999, 12.

2. *Kai Tahu* is the principal Maori *iwi* (tribe) of the southern islands of New Zealand.

3. This collaborative project is developed as a result of the Memorandum of Understanding between Otago Polytechnic and *Rūnaka* (tribal councils). The SimPā team has immensely enjoyed working with the following individual *Rūnaka,* and the committee of the combined *Rūnaka* of *Arai te Uru: Te Rūnanga o ōtākKāti Huirapa ki Rūnanga ki Puketeraki; Te Runanga o Moeraki; Hokonui Rūnaka* Incorporated Society. We would also especially like to acknowledge the late Kelly Davis and his partner Evelyn Cook from one of the key partners, *Te Matauranga Putaiao* Trust. Kelly died suddenly as the project got underway, but was a key figure in the project and is sadly missed. *Te Rūnanga o Ngāi Tahu* head office has also been very helpful in the SimPā project. We are grateful for the assistance of (then CEO) Tahu Potiki and especially the GIS expertise of Jeremy King. We have also enjoyed working with Ngāti Awa, facilitated through *Te Whare Wānanga o Awanuiārangi,* and thank Professor Graeme Smith and especially Associate Professor Mark Laws and their staff. The SimPā team has evolved over time. The authors are grateful for the support of Paul Admiraal, Jenny Aimers, Leigh Blackall, Amber Bridgeman, Justine Camp, Sunshine Connelly, Mark Crook, Rachel Dibble, Dougie Ditford, Evelyn Cook, Gwyn John, Marlene McDonald, Karen Love, Alistair Regan, Thomi Richards Lesley Smith, Dana Te Kanawa, Andy Williamson, Jeanette Wikaira, Vicky Wilson and members of the Kōmiti Kāwanataka.

4. Genealogical ties of people, of places of bodies of knowledge or knowledge to resources and seasons and food gathering. Each of these things and all things on the earth have a whakapapa and we have connections with them.

5. *Tā moko* (often referred to as tattoos) in the cultural context are as unique to the person as is their thumbprint and usually decided on by either the *tā moko* artist or *whānau* (family) member(s). Tattoos on the other hand are often given to many people based on choice or a liking for the pattern.

6. Stories of connections and connected up stories

7. Old or ancestral stores, pre-histories, but **never myth.**

8. These stories and three birds carry all forms of communication. We need to continue to learn from the messages and the news they bring.

9. Please see the Otago Polytechnic Web site for the Memorandum of Understanding.

10. Note: SimPā is shorthand for the whole project; GamePā refers to the developed game for each individual *Rūnaka* (tribal council).

11. See The *Ngāi Tahu* Vision for 2025, 17. http://www.ngaitahu.iwi.nz/Publications/Governance/NgaiTahu_2025.pdf.

12. *Tikaka* ensures appropriate cultural and interpersonal behaviors are adhered to.

13. Ancient histories based on pre-historical knowledge handed down as oral recitations of our land- and seascapes and *tūpuna kōrero* (ancestral deeds, and the doers of these deeds)

14. A building and associated complex which usually includes a *wharenui* (large building in which people learn, eat, and sleep); often ornately carved but not always; has a cemetery as part of the complex and is used for *hui* (gatherings) such as learning, celebrations, graduations, and *taki aue* (funeral rites and burials).

15. *Wānaka* or *wānanga* in the northern dialect may be sessions of intense learning on the traditional Māori meeting house (*marae*); or can be a place of learn-

ing somewhat like a University or Polytechnic. They teach either about rituals or *iwi/* Māori-based knowledge in a specific manner.

16. When the project was offered for view at the *Kai Tahu* annual gathering (*hui-a tau*), the group of young people were very interested in the end result but not so interested in how one might learn how to produce such a resource.

17. This was also protected with the formation of the Māori Intellectual Policy created and adhered to by Otago Polytechnic.

18. These are old incantations somewhat like, but also very different from, Christian prayers.

19. The brother carver who was making the carving himself died unexpectedly so his master carver finally completed the work and incorporated the original *kai whakairo* (carver) into the work to honor all three.

20. When Khyla went to London before and after presenting this paper at Norwich, she was able to engage with, report upon, and exchange stories on SimPā and its handover.

21. The space out in front of the *wharenui* or meeting house where rituals of encounter, called *pōwhiri*, occur.

22. Whilst *tikaka* may be very adaptable, *kawa* is grounded in protocols which never alter and which govern the order of this. One way of explaining this and placing it in a context we all understand would be the dress codes and expected attire we are asked to wear at formal dinners, for which there is a shared expectation that such a code will not be breached.

23. These are images of those true places of *Kai Tahu ki Otago* and those to which so many connect and long for at times. See images below text.

24. For further details of the project, please contact Dr Khyla Russell, Otago Polytechnic. http://www.otagopolytechnic.ac.nz/index.php?id=504&clientref=1029.

Contributors

Ewelina Bańka is an Assistant Professor at the John Paul II Catholic University of Lublin, Poland. Her PhD dissertation was on the representations of Indian Country in the works of Gerald Vizenor, Leslie Marmon Silko, and Sherman Alexie. Her research and teaching interests include postcolonial theory, indigenous literatures of the Americas, and border fictions. She was the recipient of the Kosciuszko Foundation grant for her project "View from the Concrete Shore: Identity and Urban Space in Contemporary Native American Literature."

Max Carocci is an anthropologist who lectures in Indigenous Arts of the Americas in the program World Arts and Artifacts, which he directs for Birkbeck College London, in joint collaboration with the British Museum. His book *Warriors of the Plains: The Arts of Native North American Warfare* (British Museum Press/McGill Queens University Press, 2012) complements a travelling exhibition on Plains Indian art that he curated for the British Museum. Among his recent publications are *Ritual and Honour* (British Museum press, 2011) and *Native American Adoption, Captivity, and Slavery in Changing Contexts* co-edited with Stephanie Pratt (Palgrave, 2011). Recently he curated *Imagi/Nations*, an exhibition about nineteenth-century Native American photography from the archives of the Royal Anthropological Institute of Great Britain and Ireland, exhibited in their headquarters in London, May–August 2012.

Jacqueline Fear-Segal is Reader in American History and Culture at the University of East Anglia, England, and has also taught at Harvard University, University College London, the Sorbonne, and Dickinson College, PA. She is co-founder and administrator [with Rebecca Tillett] of the Native Studies Research Network UK, which is based in the School of American Studies at the University of East Anglia. The author of numerous publications on Native Studies topics, her book, *White Man's Club: Schools, Race, and the Struggle of Indian Acculturation* (University of Nebraska Press, 2007), won the European American Studies Network prize in 2008. She is currently completing a study of photographs of students attending the Carlisle Indian Industrial School in the late nineteenth and early twentieth centuries: *Shadow Catchers at the Indian School: Photographic encounters and reclamations, 1879 to the present.*

Hsinya Huang is Professor of American and Comparative Literature and Dean of the College of Arts and Humanities at National Sun Yat-Sen University, Taiwan. In addition to numerous articles, her book publications include *(De) Colonizing the Body: Disease, Empire, and (Alter)Native Medicine in Contemporary Native American Women's Writings* (2004), *Lesbigay Literature in Modern English Tradition* (2008) and *Huikan beimei yuanzhumin wenxue: Duoyuan wenhua de shengsi* [*Native North American Literatures: Reflections on Multiculturalism*] (co-edited, 2009), the first Chinese essay collection on Native North American literatures. She is Editor-in-chief of *Review of English and American Literature* and *Sun Yat-sen Journal of Humanities,* and also is editing the English translation of *The History of Taiwanese Indigenous Literatures* and essay volumes, *Aspects of Transnational and Indigenous Cultures* and *Ocean and Ecology in the Trans-Pacific Context.* Her current research project focuses on Trans-Pacific Indigenous literatures.

Carolyn Kastner is the Curator of the Georgia O'Keeffe Museum in Santa Fe, New Mexico. She earned her PhD in American Art History at Stanford University in 1999. Her research, publications, and curatorial projects are focused on the diversity of American modernism. Her most recent research will be published as the monograph *Jaune Quick-to-See Smith: An American Modernist* (2013).

Marion Maar is a medical anthropologist and Associate Professor at the Northern Ontario School of Medicine (NOSM) in Canada. Her teaching and research focus is in the area of chronic illness prevention in Aboriginal communities, culturally competent health care, and child health applications. Specializing in culture and health, she was closely involved in the development of NOSM's new northern and rural focused medical curriculum. She is a faculty coordinator for a unique one month, mandatory cultural emersion experience for NOSM medical students in Aboriginal communities. She has lead and collaborated on many research projects, including Aboriginal research ethics, diabetes care and prevention, mental health, and health services research.

Darrel Manitowabi (Anishinaabe) is an Assistant Professor in the School of Native Human Services at Laurentian University, and holds a cross-appointment with the Northern Ontario School of Medicine in Sudbury, Ontario, Canada. He recently completed research on the impact of socioeconomic interventions on Aboriginal well-being and has published articles on Indigenous issues such as casinos, Anishinaabe ethnohistory, rural-urban movements, diabetes, and traditional medicine.

Samuel Mann is Professor of Information Technology at Otago Polytechnic where he has worked since 1997, including five years as Head of Department. He has published over 150 conference and journal papers in the fields of augmented experiences, sustainability, and computer education; and has been responsible for the development of Education for Sustainability at Otago Polytechnic. He is convenor of Sustainability in Tertiary Education in New Zealand (STENZ).

Murielle Nagy has been director and editor of the journal *Études/Inuit/Studies* since 2002 and adjunct professor at the department of anthropology of Université Laval since 2011. She is also a consultant in anthropology and a researcher at the CIÉRA (Centre interuniversitaire d'études et de recherches autochtones). She received an MA in archaeology from Simon Fraser University (1988) and a PhD in anthropology from the University of Alberta (1997). From 1990 to 2000, she coordinated three major oral history projects for the Inuvialuit of the western Canadian Arctic. She was awarded post-doctoral fellowships to work on the Oblate missionary Émile Petitot.

Suzanne Owen is a lecturer in Theology and Religious Studies at Leeds Trinity University, England, and primarily researches indigenous religions and their categorization. She is also co-chair of the Indigenous Religious Traditions Group at the American Academy of Religion. Her monograph, *The Appropriation of Native American Spirituality*, was published by Continuum (2008), and an article "Production of Sacred Space in the Mi'kmaq Powwow" appeared in *Diskus* 11 (2010). She is currently preparing a monograph on contemporary Druidry for Continuum.

Stephanie Pratt [Dakota] is Associate Professor of Art History (Reader) at Plymouth University. Her research has mostly concentrated on the visual representation of Native Americans in European and American art from the period c. 1600 to the end of the nineteenth century. Her monograph, *American Indians in British Art*, 1700–1840 (Oklahoma University Press) appeared in 2005. Recently, her work looks at how Native American arts and cultures have been represented in Western museums and galleries, and she is developing a book-length study of Anglo and American collections of Native American ethnographica, 1700 to 1950. She was principal curator for the exhibition "The American Indian portraits of George Catlin," at the National Portrait Gallery, London, 2013.

Carter Revard [Osage], was born and grew up on the Osage Reservation in Oklahoma. In 1952 his grandmother and the Osage elders gave him his Osage name and he won a Rhodes Scholarship to Oxford University (BA 1954), in 1959 earned a PhD 'at Yale University, then taught at Amherst College, Washington University, and (as visiting professor) at the University of Oklahoma and the University of Tulsa before retiring in 1997. His essays on medieval literature and manuscript studies, linguistics, Milton, and American Indian literature have appeared in scholarly journals. His books include: *Ponca War Dancers* (1980); *An Eagle Nation* (1993); *Family Matters, Tribal Affairs* (1998); *Winning the Dust Bowl* (2001); and *How The Songs Come Down* (2005). More recent stories and poems appear in *Stand* (2008), *Sentence* (2009), *Yellow Medicine Review* (2010), *SING: Poetry from the Indigenous Americas* (2011), and he hopes presently to publish a new collection of poems, *From the Extinct Volcano, a Bird of Paradise.*

Khyla Russell BA (Massey), PGDA (Otago), PhD (Otago) is Kaitohutohu and full professor at Otago Polytechnic (OP), New Zealand, overseeing the Treaty of Waitangi implementation across the organization, facilitated through the Māori Strategic Framework, and including leading Māori research or research specific to Māori within OP. She has responsibility for the facilitation of relationship-building (established by the Memorandum of Understanding) between Otago Polytechnic, the Arai-Te-Uru Papatipu Runaka, the wider Māori Community and other Māori tertiary providers; and she also works with Te Tapuae o Rehua Tertiary Company. Khyla is co-author of *Te Ara Tika—Guidelines for Māori Research Ethics.* She is involved in a number of collaborative research clusters and undertakes post-doctoral supervision. Her whakapapa is Kāi Tahu, Kāti Mamoe, Waitaha and Rapuwai and Hawea on te taha Māori and Polish (from Gdansk) and Northern Ireland on te taha Tauiwi.

Lynette Russell is an Australian Research Council Professorial Fellow (2011–2016) and Director of the Monash Indigenous Centre. She is an anthropological historian and author or editor of eight books, including: *A Little Bird Told Me* (2002); *Savage Imaginings (2001)*; and *Appropriated Pasts: Indigenous Peoples and the Colonial Culture of Archaeology* (co-authored with Ian McNiven, 2005).

Rebecca Tillett is a Senior Lecturer in American Literature and Culture at the University of East Anglia, England. She is co-founder and administrator [with Jacqueline Fear-Segal] of the Native Studies Research Network UK, which is based in the School of American Studies at the University of East Anglia. She has published on numerous Native Studies topics, including *Contemporary*

Native American Literature (2007). She is currently completing a study of representations of community environmental activism in the contemporary literatures of the American Southwest.

Joanna Ziarkowska is an Assistant Professor at the University of Warsaw, Poland. In 2008 she completed her PhD dissertation on the works of Leslie Marmon Silko and Maxine Hong Kingston. She has published several articles on Native American literature, and is the co-editor of *In Other Words: Dialogizing Postcoloniality, Race, and Ethnicity* (2012). She is currently working on a study exploring the function of medical discourse in Native American literature.

Index

Note: Illustrations, located after page 126 in the photo gallery, are indicated by figure number.

197